T0335737

EARTH & SPACE
SCIENCE FOR EVERYBODY

Lasse A. Kivioja

Trafford
PUBLISHING®

Order this book online at www.trafford.com
or email orders@trafford.com

Most Trafford titles are also available at major online book retailers.

© Copyright 2009 Lasse A. Kivioja.
All rights reserved. No part of this publication may be reproduced, stored in a retrieval system, or transmitted, in any form or by any means, electronic, mechanical, photocopying, recording, or otherwise, without the written prior permission of the author.

Printed in Victoria, BC, Canada.

ISBN: 978-1-4251-1975-1 (sc)

ISBN: 978-1-4269-9804-1 (hc)

ISBN: 978-1-4269-9802-7 (e-book)

We at Trafford believe that it is the responsibility of us all, as both individuals and corporations, to make choices that are environmentally and socially sound. You, in turn, are supporting this responsible conduct each time you purchase a Trafford book, or make use of our publishing services. To find out how you are helping, please visit www.trafford.com/responsiblepublishing.html

Our mission is to efficiently provide the world's finest, most comprehensive book publishing service, enabling every author to experience success. To find out how to publish your book, your way, and have it available worldwide, visit us online at www.trafford.com

www.trafford.com

North America & international
toll-free: 1 888 232 4444 (USA & Canada)
phone: 250 383 6864 ♦ fax: 812 355 4082

ACKNOWLEDGEMENTS

I owe a great debt of gratitude to many people whose talents, energy and support have made this a better book.

I thank Emily, my wife, who has given great patience, support and encouragement as I worked on this book. Her questions lead to clarification and better explanations of many items.

I am grateful to Edwin D. Posey, Professor Emeritus, for his proofreading and many suggestions.

The untiring proofreading and many suggestions for formatting and general improvements of the text by Dinah Hackerd have been greatly appreciated. I also appreciate Judy Haan for her proofreading contributions.

Many discussions on different subject matters with Bill Huston, Quil Standiford and my brother, Heikki Kivioja, have contributed to the contents of this book.

ABOUT THE AUTHOR

Lasse A. Kivioja was born in Finland and immigrated to the United States in 1955. During his studies for his Master of Science degree in Physics in The University at Helsinki he worked part-time at the Finnish Geodetic Institute. Specializing in Earth's gravity, he was an instructor and received his Ph.D. degree from the Department of Geodetic Science at The Ohio State University in Columbus, Ohio while working there at The Mapping and Charting Laboratory. He is a Professor Emeritus from Purdue University, West Lafayette, Indiana, where he lectured and conducted research in Geodetic Sciences for 26.5 years and published several refereed articles including:

In *Bulletin Geodesique*, he published a new exact non-iterative mathematical method for computing astro-latitudes and astro-longitudes for Astrolabe observations.

In *Bulletin Geodesique* and in *Surveying and Mapping* he published a method for computing coordinates and azimuths for any 'way-points' and any 'end-points' in GPS positioning, solving the two Main Problems of Geometric Geodesy (Direct and Inverse Problems) by very precise computer integration using the original differential formulas for all geodetic line elements on the surface of any Earth Ellipsoid besting all older methods. Famous mathematicians spent some time solving these two elliptical integration problems. Among them are: Clairaut 1713-1765, Lagrange 1736-1813, Laplace 1749-1827, Legendre 1752-1833, Gauss 1777-1855, Bessel 1784-1846, Jordan 1838-1922 and Helmert 1843-1917. None of these famous men had electronic calculators.

Published in *Bulletin Geodesique* studies of world sea-level variations influenced by the melting of land-supported ice masses.

Consultant at National Geodetic Survey, Fredericksburg, Virginia. Developed and published in *Surveying and Mapping* methods of improving observational accuracies of first order theodolites. Made an autocollimation addition to a first order theodolite to account for its axis wobbles.

Consultant at USAF Geodetic Survey Squadron, Cheyenne, Wyoming. Developed and published an improved method for astronomical azimuth observations in *Bulletin Geodesique* increasing achievable accuracies in the use of theodolites and leveling instruments using Mercury Leveling with autocollimation methods. Many first order theodolites were calibrated to account for small inherent systematic errors.

Consultant at Argonne National Laboratories in Argonne, Illinois. Developed and published a new method suitable for leveling the 1104-meter long ring with about 200 supports to a few micron (0.001 millimeter) accuracies in the Advanced Photon Source.

He has a United States Patent on Mercury Leveling Instruments.

TABLE OF CONTENTS

1. **INTRODUCTION**   **13**

2. **AVAILABLE BIBLIOGRAPHY ON THE INTERNET**   **18**
   2.1. INTERNET SITES, OR THE TRADITIONAL PRINTED BIBLIOGRAPHY   18
   2.2. USEFUL INTERNET SITES FOR EARTH SCIENCES   20

3. **AREAS OF EARTH SCIENCES COVERED IN THIS BOOK**   **24**

4. **FROM WHERE AND HOW DID ALL THIS COME ABOUT?**   **27**

5. **BASIC UNITS OF LENGTH**   **28**
   5.1. UNDERSTANDABLE UNITS FOR ALL ASTRONOMICAL DISTANCES   31
   5.2. HOW LONG IS THE 14 BILLION, (14 x 10E9) LIGHT-YEAR DISTANCE?   33

6. **DIRECTIONS/DISTANCES TO STARS AND GALAXIES**   **37**

7. **STAR CONSTELLATIONS**   **38**

8. **THE SUN'S LOCATION IN THE MILKY WAY GALAXY**   **40**
   8.1. PLANET EARTH IN THE UNIVERSE   41

9. **POSSIBLE ALIEN VISITS TO EARTH**   **44**

10. **COMMON VELOCITIES IN SPACE**   **46**

11. **EARTH AS A SPHERE**   **49**
    11.1. THE DIRECT PROBLEM IN COMPUTING EARTHLY DISTANCES   50
    11.2. THE INVERSE PROBLEM IN COMPUTING EARTHLY DISTANCES   51
    11.3. GENERAL SHAPE OF THE EARTH   53
    11.4. ROOM-SIZE MINIATURE BALLOON MODEL OF THIS EARTH   54
    11.5. EARTH'S ROOM-SIZE BALLOON MODEL AND THE ACTUAL EARTH   55
    11.6. ROOM-SIZE BALLOON MODEL AND THE MOON ASTRONAUTS   56

12. **MINIATURE MODELS OF THE ENTIRE UNIVERSE**   **58**
    12.1. MINIATURE MODEL NUMBER ONE OF THE KNOWN UNIVERSE   58
    12.2. LOCATION OF PLANET EARTH IN THIS MODEL   61

12.3. Miniature Model Number Two of the Universe   62

12.4. Miniature Overview Model Number Three of the Universe   62

12.5. Where are we in this Universe   63

13. EARTH'S ORBIT AROUND 'OUR' SUN   68

14. CREATOR, GOD, SUPREME BEING   73

15. NUMBER OF GALAXIES AND STARS IN THE UNIVERSE   74

16. THERE IS NOTHING IN THE KNOWN UNIVERSE LIKE THIS EARTH   77

17. THIS SOLAR SYSTEM IN THE MILKY WAY GALAXY   81

17.1. Distance Unit Parsec   81

18. ANGULAR UNITS OF MEASUREMENTS   83

18.1. Radians   83

18.2. Degrees, Minutes and Seconds   83

18.3. Hours, Minutes and Seconds   83

19. SUN AND ITS PLANETS   85

20. COMPARING THE GRAVITATIONS OF THE SUN, EARTH AND MOON   87

21. INCOMING METEOR PARTICLES AND DUST TO EARTH   89

21.1. Incoming Asteroids and Debris to Earth's Vicinity   95

21.2. Comet Tempel-Tuttle's Debris and Leonid Meteor Showers   97

21.3. Asteroids from Interstellar and Intergalactic Space   101

21.4. Annual Meteor Showers   104

21.5. Comets and Large Asteroids   106

21.6. Asteroids/Comets/Debris Approaching 'Our' Moon   109

21.7. Asteroids and Comets Hitting 'Our' Moon   113

21.8. Deadly Meteor Showers on the Moon   114

21.9. Asteroids/Comets Hitting Planet Mars, Other Planets and Their Moons   116

21.10. Asteroids/Comets/Debris Coming to the Vicinity of the Sun   117

22. LIVING CONDITIONS ON SOLAR PLANETS AND MOONS 119
22.1. MERCURY 119
22.2. VENUS 120
22.3. EARTH IS THE ONLY KNOWN GOD'S 'GARDEN SPOT' IN THE UNIVERSE 121
22.4. MARS 121
22.5. JUPITER 122
22.6. SATURN 123
22.7. URANUS 123
22.8. NEPTUNE 123
22.9. PLUTO 123

23. SIZE AND SHAPE OF THE MILKY WAY GALAXY 125

24. WATER IS ESSENTIAL FOR LIFE ON EARTH 129

25. SEARCH FOR LIFE ELSEWHERE IN THE UNIVERSE 132
25.1. POSSIBLE LIFE ON INNER AND OUTER SOLAR PLANETS 133
25.2. REASONS WHY OTHER SOLAR PLANETS/MOONS ARE LIFELESS 135

26. WE WILL NEVER MEET ALIENS FROM EXTRA SOLAR PLANETS 143

27. KEPLERIAN ORBITS 148

28. ORIGINAL BUILDING BLOCKS OF ALL MATERIALS 151
28.1. ORIGINS OF ATOMS AND MOLECULES 152
28.2. EVOLUTION/CREATION, OWNERSHIP OF THE PLANET EARTH 153
28.3. DID GOD DO IT ALL? WHAT IS CONTROLLING OUR SUN 154
28.4. IF NOT GOD, WHO, OR WHAT DID IT ALL 155
28.5. UNANSWERABLE ETERNAL QUESTIONS 156

29. SOME DETAILS OF EARTH'S MOTIONS 158
29.1. DAILY ROTATION OF EARTH 160
29.2. EARTH'S ANNUAL REVOLUTION (JOURNEY) AROUND THE SUN 160
29.3. RECAPPING EARTH'S MOTIONS WE SHARE EVERY DAY 162
29.4. EXAMPLES OF SOME SLOWER EARTHLY SPEEDS 165

30. ORBITS OF SOLAR PLANETS   168

31. TIME KEEPING, BASIC UNITS OF TIME   170
     31.1. SIDEREAL TIME   173

32. COMPASS DIRECTIONS, AZIMUTHS AND BEARINGS   176

33. GEODETIC (= GEOGRAPHIC) COORDINATES   179
     33.1. LATITUDES ON EARTH   179
     33.2. LONGITUDES ON EARTH   179
     33.3. TOPOGRAPHIC ELEVATIONS ABOVE THE MEAN SEA LEVEL   180
     33.4. CELESTIAL METHODS FOR POSITIONING   181
     33.5. GPS, THE GLOBAL POSITIONING SYSTEM   183
     33.6. ASTRONOMICAL POSITIONING   185

34. CELESTIAL COORDINATES, DECLINATION AND RIGHT ASCENSIONS   187
     34.1. DECLINATIONS   188
     34.2. RIGHT ASCENSIONS   189
     34.3. PROPER MOTION   191
     34.4. PRECESSION OF EARTH   191
     34.5. THE 23.5 DEGREE TILT OF EARTH'S SPIN AXIS   193
     34.6. NUTATION   194
     34.7. CHANDLER'S WOBBLE   195

35. EARTH'S CRUST, MANTLE AND CORE IN THE BALLOON MODEL   197

36. GRAVITY, GRAVITATION, ACCELERATIONS AND DECELERATIONS   198

37. MOON'S GRAVITY   203

38. SUN'S GRAVITY   205

39. EXAMPLES OF HIGH ACCELERATIONS   208

40. BACK TO EARTH'S GRAVITY   211
     40.1. EARTH'S GRAVITY AND CENTRIFUGAL ACCELERATION   212

40.2. Earth's Gravitational Pull has Deformed 'our' Moon 213

40.3. Direction of Gravity, Zenith (= Up) and Nadir (= Down 215

40.4. Earth Is Round due to its Gravitation 217

41. WORLD ATHLETIC RECORDS AND GRAVITY 221

41.1. Javelin Throw 223

42. GRAVITY MEASUREMENTS 225

43. GRAVITATIONAL PULLS ON THE MOON 231

43.1. More on Physical Meanings of Accelerations 232

43.2. Gravity Anomalies 233

44. GEOID (= MEAN SEA LEVEL) UNDULATIONS 235

44.1. Plumb Line Deflections Affect Astronomical Coordinates 236

44.2. Elevations and Depths are Measured from the Geoid 238

45. EARTH'S DETAILED SIZE AND SHAPE 243

45.1. Earth's Shape in the Room Size Balloon Model 245

46. TRAVELING ON THE ROUND EARTH 247

47. TIDES 251

47.1. Lunar and Solar Tidal Effects on Rotating Earth 251

47.2. Where is the Barycenter of the Earth-Moon System Located 257

48. THE SPACE AROUND US 259

48.1. No One Knows Physical Locations of Heaven and Hell 262

49. ENCOUNTERED VELOCITIES IN THE UNIVERSE 264

50. MORE ABOUT EARTH'S SOLAR ORBIT 267

51. TIME 275

51.1. What Time is it? What is the Origin of our Time Keeping? 275

51.2. Universal Coordinated Time (UTC) and Leap Seconds 276

51.3. Sidereal Time = Star Time 277

51.4. Leap Years by Julian and Gregorian Calendars 278

51.5. Sidereal Years and Mean Solar Years 280

51.6. SUNDIALS   280
51.7. REASONS FOR THE EXISTENCE OF THE EQUATION OF TIME   282

52. PLANET EARTH, OUR SWEET HOME   284
52.1. REASONS WHY THIS EARTH IS A HABITABLE PLACE FOR HUMANS   285
52.2. COMPARING EARTH TO MERCURY   287
52.3. COMPARING EARTH TO VENUS   288
52.4. COMPARING EARTH TO MARS   289
52.5. ATMOSPHERIC MODERATION OF TEMPERATURES ON MARS   290
52.6. JUPITER, SATURN, URANUS AND PLUTO ARE NOT LIVABLE PLACES   290

53. THE SOLAR SYSTEM IN THE BALLOON MODEL   292

54. GLOBAL TEMPERATURE MODERATIONS BY
THE ATMOSPHERE   294
54.1. GLACIER ON KILIMANJARO MOUNTAIN   296
54.2. A PARTIAL LIST OF THE BENEFITS OF OUR ATMOSPHERE   297

55. EARTH'S INTERIOR: CRUST, MANTLE AND CORE   299
55.1. EARTH'S INTERIOR IN THE 1:5,000,000 SCALE BALLOON MODEL   301
55.2. ARCHIMEDES' LAW OF BUOYANCY   301
55.3. ISOSTASY = ARCHIMEDES' PRINCIPLE FOR THE
INNARDS OF THE EARTH   304

56. GROUND MOVEMENTS UP, DOWN AND SIDEWAYS   309

57. EARTH'S OCEANS   313
57.1. OCEAN CURRENTS   315

58. EARTH'S ATMOSPHERE   318
58.1. ATMOSPHERE HAS MANY FUNCTIONS   321
58.2. FURTHER BENEFITS OF EARTH'S ATMOSPHERE   323

59. GLOBAL WARMING AND COOLING   325

60. INDEX   328

1. INTRODUCTION

Most people are interested in better understanding the realities of the actual world around, above and under us. That is the purpose of this book.

This book is written in 'plain English' so that it is easy to read and understand. The book puts this planet Earth, our Solar System, the Milky Way Galaxy and all other known galaxies almost 'into a bottle' to be 'seen' by all. The described miniature 3-dimensional model is literally *The Real Overview Model* of the entire known universe. It is small enough to fit in many backyards.

Many items presented in this book is new information to persons whose area of expertise is not in the areas of Earth Sciences, Astronomy or in Space Sciences.

Foremost, this book is an **overview of general, factual and up-to-date scientific information** about this wonderful, unique planet Earth from the top of its atmosphere down to its center core.

This Earth is truly the 'Garden Spot' in our solar system and even in the whole presently known universe. This Earth has water, food and breathable air. No liquid water, food or breathable air has been found to exist anywhere else in the universe. **Many of us take the thousands upon thousands 'things' available to us here on Earth for granted.** Many of those 'things' are actually the very 'things' making our lives possible.

This Earth is 'it'! Love it! Don't mess it up! All of us are fortunate to be able to live our life span on this Earth. Try to improve 'our' Earth the little you can and keep it neat!

Where are we and where is the Earth going?

The instantaneous location of planet Earth is at some spot in its annual, almost circular (elliptical) orbit around the Sun at a distance of approximately **500 light-**

seconds = 1 AU (Astronomical Unit) = 149,597,871 km = 92,955,622 miles. (Note that light-seconds, light-minutes, light-hours and light-years are distances). **Earth travels in its solar orbit at an average speed of 30 kilometers = 18.6 miles per second**, or 30 times as fast as a bullet from a military rifle.

The Earth orbit is in the plane of the ecliptic = Ecliptica = Zodiac = the path in the skies our Sun seems to travel every year among its background stars as seen from the Earth. Our orbital plane is a very steady plane year after year.

This Earth is one of our Sun's nine (eight, 2006) orbiting planets. Planet Mercury is nearest to the Sun, orbiting at a distance of 190 light-seconds from the Sun.

Pluto is the most distant of the nine major solar orbiters. Its distance from the Sun is 19,700 light-seconds = 5.5 light-hours = 39.5 AU (Astronomical Units).

The so-called Kuiper belt and the Oort Cloud, with millions of asteroids, comets, boulders, gravel and dust clouds, orbit our Sun beyond Pluto's orbit reaching to a distance of at least 50 Astronomical Units.

Most of the more or less elliptical orbits of all these solar orbiters are in almost the same plane with the Earth's orbit (Ecliptica, = Zodiac) making this solar system a disk shaped entity.

The following Internet sites have good diagrams and pictures of our solar system:

http://www.nineplanets.org/overview.html
http://solarsystem.nasa.gov/planets/profile.cfm?Object=OortCloud
http://www.astronomynotes.com/solfluf/s8.htm

Beyond this solar system, at a distance of approximately 260,000 Astronomical Units (AU) = 4.2 light-years, is our nearest star, Proxima Centauri.

Further out, the physical location of our Sun with its entire planetary system, together with our 'nearby' stars, **is in one of the pinwheel arms of our disk-shaped home galaxy called the Milky Way Galaxy**, consisting of more than 100 billion stars, boulders, gravel and clouds of dust. Our Sun is one Milky Way star. All Milky Way stars orbit the Milky Way's deadly center. It is deadly due to its intense radiation. In addition, that central part is crowded with many fast-moving dangerous objects.

Our solar system is orbiting the Milky Way center at a safe distance of approximately 25,000 to 28,000 light-years. **The diameter of the Milky Way disk is** approximately **100,000 light-years**.

This Milky Way Galaxy is just one galaxy among 'billions and billions' of other galaxies and clusters of galaxies. The galaxies are scattered around all the way to the edges of the known universe in all directions from here.

The known universe is believed to be spherical or of some other unknown shape all around us with its 13.7 billion = 13,700,000,000 light-year long radius. All galaxies are moving in their own directions. Everything moves in space. Some galaxies are in the process of colliding; some are a few million light-years apart.

The total number of galaxies in the universe is unknown. The published total number of galaxies ranges from 125 billion to 3,000 billion.

The nearest galaxy similar to this Milky Way Galaxy is the Andromeda Galaxy, approximately 3 million light-years away from us. Andromeda and Milky Way Galaxies are on a collision course decreasing their mutual separation by approximately one light-year in one thousand years. Therefore, there is still approximately 2,999,999 years to go before the looming collision.

The much-unknown detailed beginning of all material things in the universe is called the Big Bang or Creation. According to the best present estimates, the Big Bang happened 13.7 billion years ago, and this solar system is estimated to be 4.6 billion years old. Many scientists believe that the necessary atoms came from star explosions in this Milky Way galaxy.

The following **mostly unknown Evolution or Creation** (take your pick!) had to have some atoms and molecules to work with, finally producing good life on this planet Earth. The much-debated personal opinions of Creation and/or Evolution cross many religious lines and beliefs. This book cannot shed any useful light on those disagreements.

Many mutations, developments and changes during the long Evolution (4.5 billion years for the Earth) after the Big Bang/Creation have produced very clever and intelligent designs/results in many respects leading to our existence on this livable planet Earth.

The author has asked a few medical doctor specialists if they could improve on the designs of the human eye, brain, ear or the production of normal human blood cells at a rate of 2.5 million per one second. Not one doctor has said that he/she could.

So far, very little is known about any other solar system anywhere in the Milky Way Galaxy, much less in other galaxies. However, it is a known scientific fact that there are no other planets suitable for human life in this solar system, nor within at least an 8 light-year distance from us.

In this book, **the Internet is a convenient extra help replacing the traditional bibliography** section for additional information when it is desired. **More up-to-date details, tables, photographs and good descriptions are available at the carefully chosen reputable Internet sites given in the text than what is possible to have in any printed book.**

The Internet sites are selected so that the first link usually gives the desirable extra scientific information for the topic in question. **All selected sites are in the Public Domain**, therefore their usage is free to everybody.

The Internet connection is not necessary to have for reading this book. However, many readers have the Internet available; those who don't can go to libraries to find the indicated and recommended sites as desired. All readers can benefit from the given Internet sites.

There are many Internet Search Engines capable of taking the reader to a desired page/site. **This book uses the Google Search Engine** for the desired Internet sites/links/addresses/pages. **All Internet sites are selected to be reputable quality sites** such as US government sites: **NASA, JPL, NIST, AGU, USGS and NOAA** and many more. Just typing those mentioned acronyms in the search engine and pressing ENTER/RETURN brings the desired information to the screen of the monitor almost immediately.

For example, when connected to the Internet browser (Explorer, Firefox, Netscape, Safari, etc.), **go to <www.google.com>**. Type the given item into Google's search box. The suggested word/words could be **"jpl". (Ignore the quotes).** The desired *underlined* links will appear usually within one second.

Click on any underlined link (word/words). The first link is usually the intended one. Many other links on the same Internet page can also be informative. The Internet addresses in this book are in **bold letters** for easier typing the indicated letters/word/words into the Google's search box. If the given link is an Internet address, type it to be your next Internet site, in which case the Google is not needed.

For example if in the text, the recommended Internet site appears as

Google: < jpl >

Then type "jpl" (without quotes) into Google's search box. The first under-lined link will be 'Jet Propulsion Laboratory'. By clicking on it, a wealth of JPL information will become available for the **Earth, Solar System, Sun, Stars and Galaxies, Planets, Asteroids, Comets and some other items**. Click on the one that interests you. After reading about one item, press GO BACK, and then select another *underlined* item to read if desired.

If pictures/photos/information of this Milky Way Galaxy, the 'nearby' galaxies' and other galaxies are of interest, use the following sites:

Google: < the milky way >
**   < list of nearest galaxies >**
**   < NOAO Image Gallery: Galaxies > Click on photos.**
http://antwrp.gsfc.nasa.gov/apod/ap070319.html

To avoid typing the recommended Google search terms and the Internet addresses listed in the text, visit the author's website **www.lakivioja.com.** From there download the 'Quick Pick Internet Links'. Then each Internet site can be accessed rapidly with one click of the mouse.

2. Available Bibliography on the Internet

The **Bibliography section** is commonly at the end of many books. The traditional printed version of the bibliography **is omitted in this book.** Instead, **the Internet can 'do the bibliography' better.** Many books, including this one, are read without giving much attention to the bibliography section.

In this book **the Internet is a great convenience and an enlargement replacing the conventional bibliography section. Recall from earlier that the Internet is not necessary for reading this book.**

2.1. Internet Sites, or the Traditional Printed Bibliography

Related references are more readily available on the Internet with more abundance and updated information than what is available in the bibliography section of a printed book. The numerous Internet addresses given in the text will bring up many details, histories, photographs, illustrations and tables more abundantly than any one book can do, no matter how extensive its bibliography is.

Internet assistance is very valuable for those readers who are interested in additional details and amplifications, as well as for the ability to make quick checks for the correctness of statements made in this book.

The Internet sites given in this book were chosen to be reliable so that the helpful links will appear among the first links by the search engine. All given sites are in public domain, and therefore they can be freely and quickly used by everyone.

However, as was already mentioned, to be able to understand and enjoy the written material presented in this book, **one does not need to be connected to**

the Internet at all reading sessions. The material covered is explained in sufficient detail for general knowledge and understanding without going into the latest high precision specialty details.

Once more: the given Internet sites are replacing only the traditional bibliography section, and then some!

Readers without Internet access, who desire additional background information, normally must go to a library anyway to find the extra-desired information. Nowadays, most libraries have computers available with Internet access and librarians who can provide helpful information as needed.

Even in libraries, specialized information in the field of Earth Sciences is generally more readily and abundantly available on the Internet than on the library shelves. In addition, the Internet is faster than trying to find a book/publication by walking among the bookshelves, which very well may not even have the desired information.

Many of the given Internet sites are regularly updated as new information becomes available. This makes it the ultimate resource for research and news. Compare this to any book with its printed bibliography section! Once a book and the bibliography section are printed, they cannot be updated without printing a new edition or an addendum.

Still, **traditional library books and periodicals** are available in libraries, and they **are very useful for many studies.** In some cases library books and periodicals have information that cannot be found on the Internet. The Internet does not have everything either. Sometimes The Library of Congress might be needed.

Google: < library of congress >

One should also keep in mind that the Internet has not been in common usage much before 1990, and that scientific information becoming available since 1950 in many respects is more abundant than all the information collected before that time. Many books in libraries are outdated, but not all of them. **Many Internet sites are updated regularly.**

However, we all stand 'on the shoulders' of our parents, teachers, numerous scientists like Erathostenes, Pythagoras, Hippocrates, Copernicus, Galileo Galilei, Kepler, Newton, Einstein and many others. By the way, just by typing the last name of one of those or other famous scientists in the **www.google.com** search box, many of their histories and accomplishments will be just a click away on the computer's monitor screen, usually within one second.

Once again: all Internet sites/addresses/links given in this book are in the public domain and therefore, they can be used freely by everybody.

Google: < **copyright term and the public domain in the united states** >

2.2. USEFUL INTERNET SITES FOR EARTH SCIENCES

Extensive, reliable information about Earth, the Solar System, the galaxies and the universe can be found at:

- NASA at: **www.nasa.gov and Google:** < **nasa earth observatory** > **http://solarscience.msfc.nasa.gov/SolarWind.shtml** **http://solarscience.msfc.nasa.gov/** **www.space.com** 'From satellites to stars'. **www.space.com/news**

- Earth Observatory. New Images every week: **http://earthobservatory.nasa.gov/Newsroom/**

- Jet Propulsion Laboratory at: www.jpl.nasa.gov, or simply at **www.jpl.com**

- National Oceanic and Atmospheric Administration (NOAA) at: **www.noaa.gov**

- U.S. Geological Survey at **www.usgs.gov**

- U.S. Government at: **www.science.gov**

- **U.S. Naval Observatory:** **http://tycho.usno.navy.mil/what.html or www.usno.gov**

- U.S. Naval Research Laboratory at: http://www.nrl.navy.mil/

- Sun – Earth environment: **www.spaceweather.com**

- HUBBLE:
  http://hubblesite.org/newscenter/archive/releases/2007/01
- American Geophysical Union (AGU) at: **www.agu.org**
- **Chandra X-Ray observatory:**
  http://www.nasa.gov/mission_pages/chandra/main/index.html
- SOHO, Solar and Heliospheric Observatory:
  http://sohowww.nascom.nasa.gov/data/realtime/eit_284/512/
- SECCHI: Sun Earth Connection Coronal and Heliospheric Investigation:
  http://secchi.nrl.navy.mil/
- STEREO-A and STEREO-B: **http://stereo.gsfc.nasa.gov/**
  http://ares.nrl.navy.mil/~wang/STEREOviewer/
- ESA, European Space Agency:
  http://www.esa.int/esaCP/SEM0GW8L6VE_index_0.html
- JAXA, **Japan Aerospace Exploration Agency:**
  http://www.isas.jaxa.jp/e/
- NAOJ, Hinode, National Astronomical Observatory of Japan:
  http://solar-b.nao.ac.jp/index_e.shtml
  http://science.nasa.gov/headlines/y2006/02nov_firstlight.htm
  http://www.virtualobservatory.org/

Orbital Astronomical telescopes:
  http://www.seds.org/~spider/oaos/oaos.html

Ground-based Astronomical Observatories:
  Mauna Kea Observatories: **http://www.ifa.hawaii.edu/mko/**
  http://dir.yahoo.com/Science/Astronomy/Research/Observatories/
  http://tdc-www.harvard.edu/mthopkins/obstours.html
  http://www.astro.caltech.edu/~pls/astronomy/observs.html

One does not need to use bold or capital letters for Internet addresses nor in the search boxes of Internet search engines. **They are in bold in this book so they will be easier to enter into the computer.** This book will use the Google search engine. Many other search engines exist.

For instance, one of the links (among the underlined items = clickables) in the mentioned US Government site, is **"Earth and Ocean Sciences"**. After clicking on it, many new links will become available with a wealth of reliable up-to-date information.

Please take a good look at some of those Internet sites given above. Click on the links that interest you. As mentioned, if you are in a library, the librarians are available for computer instructions. At the next visit you won't need much help.

American Geophysical Union scientific research covers the following areas in Earth Sciences:

Atmospheric Sciences
Biogeosciences
Geodesy
Geomagnetism and Paleomagnetism
Hydrology
Ocean Sciences
Planetary Sciences
Seismology
Space Physics and Aeronomy
Tectonophysics
Volcanology

Definitions for those AGU Earth Science areas are available by just typing the desired word in the **www.google.com** search box, which is faster than using a traditional dictionary.

American Geophysical Union has over 41,000 members (2004) from most countries in the world. Various AGU journals publish new information every week/month.

The following listing gives a good idea of the areas of study in the field of Geodesy in the Earth Sciences for one AGU Meeting.

'"Abstract submissions are being accepted for the 2006 AGU Fall Meeting, 11-15 December 2006, held in San Francisco, CA. The complete list of sessions for your section is listed below for your convenience. Both the session code and title are included.

The deadline for abstract submissions is 7 September 2006. For more information, visit www.agu.org/meetings/fm06/

G01	Geodesy General Contributions
G02	Enhanced Geophysics by Combinations of Independent Geodetic Measurements
G03	Geodetic Studies in Regions of Extension
G04	Nonsecular Changes and Variability in Regional Sea Level
G05	Coastal Subsidence, Sea Level Rise, and Gulf Coast Hazards
G06	Postseismic Deformation: Measurements, Mechanisms, and Consequences
G07	Transient Strain Accumulation Across Continental Fault Systems and Implications for Time-Dependent Seismic Hazard
G08	Geodetic Laser Scanning Methods and Applications
G09	InSAR Science Results and Recommendations for Future Missions
G10	Multidisciplinary Results and Applications From the GRACE Mission
G11	Plates, Microplates, and Blocks
G12	Planetary Geodesy
G13	Earth Rotation/Polar Motion at Rapid Timescales
G14	Plate Motion, Continental Deformation, Space Geodesy, Seafloor Spreading, and Transform Azimuths
G15	Geodetic and Geophysical Applications of the International Terrestrial Reference Frame: Current Status and Future Needs
G16	Glacial Isostatic Adjustment and Its Role in Secular Trends in Gravity, Hydrology, Sea Level and Cryosphere: New Observations and Constraints
G17	Volcano Geodesy"

3. Areas of Earth Sciences Covered in this Book

D ue to the variety and the large amount of human knowledge and re-search in the Earth Sciences, **any one book on any particular subject matter can cover only a certain part of available information**. Inter-net supplementation can be very informative and helpful.

This book brings out known interesting facts about this planet Earth and about 'our' Solar System, 'our' Milky Way Galaxy and the surrounding universe **in an easy-to-understand manner.**

The Internet addresses given in the text have been carefully selected to provide fast access to the links further describing the subject matter in question. The links from Google appear in a fraction of a second. Clicking on a link usually brings the text and/or pictures on the monitor screen within a few seconds.

Human minds were made to be inquisitive. A normal person wants to know and learn more about numerous 'things' of interest in her/his life. It is quite natural and normal to be interested in the many aspects of our families, jobs, homes, professions and daily living here on this Earth where we spend our earthly lives. We have a tendency to take the most basic things in life on Earth for granted. **Earth and all life on it is a miracle in many ways**. We all are the recipients and lucky stewards.

There will always be much more to learn and some unanswered questions will remain unanswered about this Earth even as new re-search uncovers some of the unknowns. Among the many 'items' around us, Earth Sciences shed light on the following areas. The purpose of this book is to 'whet appetites' in Earth Sciences for general knowledge, under-standing and appreciation of what is at our disposal here on Earth every second of our lives.

Most people are interested in:

Villages/
Towns/
Cities/
Surrounding lands/
Our country/
The continents/
The oceans/
The atmosphere/
Size and shape of this Earth/
Latitudes, longitudes, elevations, compass direction/
GPS/
Earth's gravitation and gravity/
Earth's interior/
The surrounding space beyond Earth's atmosphere/
Asteroids, comets, meteorites/
Earth's artificial satellites/
Earth's Moon/
This solar system with its Sun, its planets and their moons/
Orbits and motions in space of Earth and other solar planets/
Time keeping/
Leap seconds/
Milky Way Galaxy/
Number of stars in Milky Way Galaxy/
Sun's and our location in Milky Way Galaxy/
Sun's life expectancy/
Earth's orbital plane and location and orientation in Milky Way Galaxy/
Nearest stars/
Nearest solar systems/
Possible Alien visits to Earth/
Nearest galaxy similar to Milky Way/
Other galaxies in the universe/
Number of galaxies and stars in the known universe/

Size of the known visible universe/
Miniature model of the whole known universe/
Estimated age and size of the known universe.

This book will give general information on some of the subject matters listed above. Of course, it is impossible to know everything, but we do know something and are learning more as time goes on.

Learning is and should be a life-long process for everyone. There is an endless discovery about this Earth, Solar System and the universe. New information is published monthly/weekly and daily on some Internet sites.

4. From Where and How did all this Come About?

This question can never be answered in complete detail. No one knows the billions and billions of exact details of the origins of the universe, the origins of this Earth or the origins of life here on Earth. **This Earth in 'our' solar system and this solar system in 'our' Milky Way home galaxy are all fine tuned in countless ways making our very existence possible. But we do know that the universe exists, this Earth exists and that we do exist here.**

Not all people believe in the existence of God, or in any kind of intelligent design in starting and developing the known universe. To suit most readers and most opinions, if desired, some other words of one's own personal preference are suggested substitutions for the word God, such as the name of **your own God, Supreme Being, Probabilities, Statistics or whatever else is preferred. The words God, Creator and the Supreme Being will be used in this book. These words carry individual interpretations by the millions!**

No human being can know, nor understand 'all of it' anyway, so all of us should feel free to have our own beliefs and informed opinions freely according to our own understandings/convictions/beliefs! Most of us yearn for more factual information based on sound science, and more of it has been forthcoming every year. **The aim of this book is to report the outlines of the latest general findings and knowledge about this Earth, 'our' Solar System and the universe. The intention/goal here is to 'tell it like it is'.**

Wherever/whenever a presented item in this book does not seem to be sufficient or completely understood from the written text, the numerous Internet sites, which are only a click or two away, can provide some clarifications, reviews and memory checks. **Many Internet sites are updated regularly to keep them current**. This cannot be said of an edition of any printed book with the traditional bibliography/reference section. This book partially updates itself!

5. Basic Units of Length

Definitions for the English length units such as one inch, one foot and one mile are based on the metric system by United States law.

Google: < nist > and for all SI (International System) units at: **http://physics.nist.gov/cuu/Units/current.html**

The meter (abbreviation, m) is the unit of length in the International System of Units, called the SI system. Inches and feet are derived from the length of one meter.

Modern definition of a meter is tied to the velocity of light in a vacuum, and it is as follows:

One meter is the distance traveled by a ray of electromagnetic (EM) energy through a vacuum in 1/299,792,458 seconds, or in 0.000 000 003 356 400 95 seconds, or in 3.335,640,952 ns (nanoseconds), or in 3.335 640 952 x 10E-9 seconds. The exponential notation 10E-9 stands for ten to power minus nine. Note that the velocity of light is exactly 299,792,458 meters per second. This kind of accuracy is needed in many precise scientific applications. This definition became possible after the atomic clocks became sufficiently accurate. Note that the length of one meter is tied to the velocity of light in a vacuum.

One of the first definitions in the days of the Metric System of one meter was the distance between two scratch-marks on a special Platinum-Iridium bar kept in Paris, France. Most countries have copies of that Platinum-Iridium meter bar. The one-meter distance on the Platinum-Iridium bars was derived from the size of Earth approximately two hundred years ago.

An act of US Congress made it official in 1866 stating: *the basic American length unit is the meter*. **It is still valid and the law of the land.** The lengths of inches, feet, yards, fathoms and statute miles are derived from the meter.

The law of the land (USA) is: *one meter is exactly = 39.37 inches.*

Google: < base unit definitions: meter >
                   < definition of the meter>
                   < the united states and the metric system >
                   < usno master clock time >
                   < nautical mile >

Other familiar length units are derived from the basic definition as follows:

From 1 inch = 1/39.37 of one meter = 25.40005080 millimeters. **The Industrial inch is rounded off to 25.4 millimeters.** It is widely used by many manufacturers.

The other familiar length units are obtained as follows:

One foot = 12 inches
One yard = 3 feet
One fathom = 6 feet
One mile = 5280 feet
One nautical mile = 1852 meters = 72,913 inches = 6076 feet. A nautical mile is the length of one minute of a meridian arc at a selected latitude. This selection is made because the length of one minute (or one degree) of a meridian arc is shorter at low latitudes than the same is at higher latitudes. An approximation can be obtained as follows: The length of the meridian arc from the Equator to either pole is 10,000 kilometers, and in angular measure, it is 90 degrees, or 5,400 minutes (90 x 60 = 5400). By dividing 10,000,000 meters by 5400, one gets 1851.85…meters, which is a good approximation to 1852 meters.

As was mentioned, initially the length of one meter was tied to the length of the meridian arc from the Equator to the North Pole.

Information about the meter can be found by typing "**platinum-iridium meter bar**" (without quotes) into **www.google.com** search box, and by clicking on some of the links on the site.

Google: **< platinum-iridium meter bar >**

Atomic clocks can define the length of the meter much more accurately than the distance between two scratch-marks on the old standard Platinum-Iridium meter bars.

Before the Metric System was chosen and adopted, there were many length units in use in various countries. The intention in selecting the length of the meter was to find the length of one meter so that the total length of the 360-degree meridian ellipse around the globe, which is almost a circle, would be 40 million meters or 40,000 kilometers (24,855 miles). Thus, **the length along the Earth's surface from Earth's Equator to the North Pole (meridian quadrant running through Paris, France) would be 10 million meters or 10,000 kilometers (6214 miles).**

The metric system is a part of the SI system of measures. Letters SI come from the French language (adjective after the noun) expressing International System of Units.

Google: < nist history of metric system > NIST stands for National Institute of Standards and Technology.

            < si system of measures >

            < nist metric information and conversions: a capsule history >

            < npl history of the length measurement >

            < metric system >

            < nist >

            < international system of units >

            < old units of length >

            < the earth based units of length >

In the French determination of the length of the meter some two hundred years ago, surveying and astronomical measurements (observations) were performed along surveying triangulation chains mainly in France, Scandinavia and Peru. Using spherical trigonometry (ellipsoidal trigonometry later) for lines on Earth's surface and data from the surveying measurements, the total length between the end points with their latitudes and longitudes of each triangulation chain could be computed in those length units, which were used at that time.

Thus, the length of the meridian quadrant from the Equator to the North Pole in the original length units became known. When that distance was divided by ten million, the length of one meter could be scratched on the Platinum-Iridium meter bar.

5.1. UNDERSTANDABLE UNITS FOR ALL ASTRONOMICAL DISTANCES

The total mass of 'our' Sun is 1.989 x 10E27 metric tons. This is approximately 333,000 times the mass of the Earth. The Sun contains 99.8% of all the mass of this Solar System including all the nine/eight planets, their moons, asteroids and comets.

The Sun's diameter is 1,390,000 kilometers (863,700 miles), or approximately 3.86 times the distance from here to the Moon.

Earth's mass is 5.972 x 10E21 metric tons, and Earth's average radius is 6,371 kilometers (3959 miles).

The Moon's mass is 7.35 x 10E19 metric tons, and the Moon's average radius is 1738 kilometers (1080 miles).

The mean density of the Sun is 1409 kilograms per cubic meter and Earth's mean density is 5515 kilograms per cubic meter. Therefore, Earth is almost four times as dense as the Sun.

Google: < anatomy of the sun >
              < the nine planets > has data on this solar system
              < solar system sizes and scales >

To be able to describe and to understand 'our' **solar system and astronomical distances**, some suitable units are needed just like millimeters, inches, feet, meters, kilometers and miles are needed in our daily lives.

The following distance units are in common use for astronomical distances:

1. One **AU = Astronomical Unit** is **Sun's average distance from Earth = 149,598,000 kilometers** = 92,956,000 miles = 500 light-seconds.

2. One **light-second** (ls) is the **distance light travels in vacuum in one second**. Its length is = 299,792.458 **kilometers = 186,283 miles**, or 7.5 times the length of Earth's Equator.

3. One **light-year** (ly) is the **distance light (electro-magnetic radiation) travels in vacuum** in one year, or in 31,557,000 seconds. It is = **9,460,500,000,000 kilometers = 5,878,500,000,000 miles = 63,240 Astronomical Units.**

4. More seldom-used astronomical distance is **one parsec** = 3.08568 x 10E13 kilometers = 4.965933 x 10E13 miles = 3.26 light-years.

After only remembering the items in 1, 2 and 3, one has a good grasp of all distances from millimeters, inches, feet, meters, kilometers and miles to light-seconds, Astronomical Units and to light-years. *Nothing else is needed to describe any distance from the smallest fraction of a millimeter all the way to billions of light-years to the edge of the known universe and even beyond.*

The following practical examples make it easy to stay on top of and to understand all numerical distances used for the cosmos. It becomes clear, whether the discussion is about earthly, lunar, solar system, Milky Way or galactic distances.

As was mentioned, in one second light travels a distance which is equal to 7.5 times the length of Earth's Equator.

The length of Earth's Equator is 0.133 light-seconds = 40,000 kilometers = 24,855 miles.

The Moon is at a distance of 1.2 light-seconds = 60 times Earth's radius from Earth.

The Sun is at a distance of 500 light-seconds = one Astronomical Unit from Earth. 1 AU = 149,598,000 km = 92,956,000 miles.

The most distant solar planet Pluto is 39.5 Astronomical Units = 19,750 light-seconds = 0.0006 light-years from the Sun. On the average, it takes a radio signal 19,750 seconds = 5 hours 29 minutes to reach Pluto from here.

The Astronomical Unit is a reasonable length unit to express solar-system distances, and the light-year is a reasonable length unit to express distances to the 'fixed' stars and galaxies.

The nearest star is approximately **4 light-years from us.**

The diameter of this Milky Way Galaxy pinwheel disk is approximately **100,000 light-years**.

The most distant visible galaxies (2007) from us are at a distance of 13.7 billion = 13,700,000,000 = 13.7x E9 light-years in any and all directions from here.

The nearest large Milky Way type spiral galaxy, the Andromeda Galaxy, is approximately 2.5 million (estimates range from 2.3 to 2.9 million) light-years from us. The diameter of the Milky Way disk is approximately 100,000 light-years and Andromeda's diameter is estimated to be 220,000 light-years (2005). Andromeda and Milky Way are just two galaxies of many billions of galaxies in the known universe. Therefore, any object closer than approximately 80,000 light-years from Earth is a part of the Milky Way Galaxy. (Note that we are approximately 28,000 light-years from the center of the almost circular Milky Way Galaxy disk, and from the Milky Way's center, it is 'only' 50,000 light-years to the Milky Way's edges in all directions along disk.)

Google: < apod index-galaxies local group >
            < list of nearest galaxies > This site lists 29 galaxies within 3 million light-years from us. Andromeda is one of them. The volume of a sphere with a radius of 3 light-years is 113 cubic light-years. Therefore in this small sample, the density of such galaxies around us is approximately 0.25 galaxies per one cubic light-year, or approximately one such galaxy in a volume of about 4 cubic light-years. The density of galaxies in the universe is not uniform.

When these few simple distances are understood, the order of magnitude of all astronomical distances becomes clear. It is that easy!

5.2. How Long Is The 14 Billion, (14 x 10E9) Light-Year Distance?

To express astronomical distances and quantities, large numbers cannot be avoided. To refresh memories if desired/needed, see the following Internet sites:

Google: < fundamental physical constants >
            < one light year >
            < exponential notation of numbers >

Recall that one light-year distance is the distance light (electromagnetic radiation) travels through vacuum in one year. The velocity of light is approximately 300,000 (precisely 299,792.458) kilometers per second (186,282.0245 miles per second). It is equivalent to traveling around Earth's Equator 7.5 times in one second.

Recall also that the mean (average) solar year has approximately 365.24219 mean solar days, and that each day has 24 hours and each hour has 3600 seconds. Using a calculator, one can find (pocket calculators of high school students will suffice) the length of one light-year from the following product: 365.24219 x 24 x 3600 x 299,792.458 kilometers = 9,460,528,000,000 kilometers = 9.460528 x 10E12 kilometers = 5,878,487,000,000 miles = 5.878487 x 10E12 miles.

Also recall that these large numbers are best written using exponents of number ten, which count the number of digits from the decimal point. The (invisible) decimal point is right after the last zero in those two previous large numbers 9,460,528,000,000 and 5,878,487,000,000. Using exponents of number ten and only three significant number accuracy, those two large numbers can be written:

So, one light-year distance is 9.46 x 10E12 kilometers = 5.88 x 10E12 miles with sufficient accuracy for general purposes. The nearest star to us is at 4.24 light-year distance.

To help all readers grasp the needed orders of magnitude of the unavoidable big numbers, the following repetitions may be instructional.

Therefore, the known universe is within a sphere with a radius of 14 billion = 1.4 x 10E10 light-years, or (1.4 x 10E10) times (9.46 x 10E12) = 1.3 x 10E23 kilometers = 8.2 x 10E22 miles all around us. It is a very large sphere, spheroid or of some other shape! If there is something outside that volume, so far only God knows about it.

Without using exponents of number ten, the 14 billion light years in kilometers and miles are written as 130,000,000,000,000,000,000,000 kilometers = 82,000,000,000,000,000,000,000 miles. Using exponents, the 14 billion light-years can easily be expressed even in millimeters just by increasing the exponent by six, because every kilometer has one million millimeters. So, it could be said that the 14 billion light-years is 1.3 x 10E29 millimeters long. Nobody is afraid of large numbers anymore, or are you?

All of us are considered to be at the center (or near to it) of this tremendously large sphere (volume) containing everything in the known universe. In other words, it is OK for us to assume and to consider being at the center (or near to it) of the whole universe! The Earth's radius of 6371 kilometers (3959 miles), and even the average distance to the Sun, which is **one Astronomical Unit: 1 AU = 149,600,000 kilometers = 92,960,000 miles = 500 light-seconds,** are insignificantly small when compared to the 14 billion light-years. Recall that one year has 'only' about 31.5 million seconds, or about that many human heart beats.

The contents inside the sphere with the 14-billion light-year radius around us are only partially known. Something new about the universe is published every week.

It is not unreasonable to assume that the Creator knows everything about this immense universe and beyond, and He has not forgotten anything about it. The Big Bang and star explosions made all the existing elementary particles/atoms/molecules/cells/life and put them to their proper places in this enormous universe. **Details of those events will remain unknown to us.** There are millions of differing opinions on that matter among world's religions.

Of course, some things are known. For instance, it is commonly believed that all natural elements heavier than oxygen were created in star explosions (supernovae) during the past few billion years. That includes the calcium in our bones and iron in our blood. Details of the origins of the explosive matter remain unknown.

Google: **< periodic table >**

                < supernova >

http://www.damtp.cam.ac.uk/user/gr/public/bb_home.html

Humans are able to study some of the wonders in the universe. **We have good reasons to be grateful for those 'few' atoms/molecules/cells that make up our very complicated living bodies, and of course, for the opportunity to live here on this marvelous and unique planet Earth.**

There is nothing else like this planet Earth at least within a 10 light-year distance. With many good reasons this Earth can be called Creator's/God's garden spot in the whole known universe. So far, no other place even resembling this planet Earth has been found anywhere in the whole universe.

However, there might be billions of earth-like planets out there even **in this Milky Way Galaxy,** but due to the insurmountably long traveling/communicating distances to those unreachable stars and their planets, they are and will remain unreachable to us. The extra solar planets (orbiting some Milky Way stars) found so far have Jupiter-size masses or bigger. Their gravities alone would be deadly for humans. Assumptions have been made that there might be Earth-size livable planets nearer to their suns, but they are still more than 10 light-years away from us!

Google: < extra-solar system planets >

                < other planetary systems

6. DIRECTIONS/DISTANCES TO STARS AND GALAXIES

The directions to the 'fixed' stars from Earth, as we see the stars, are said to be from anywhere on Earth to the **surrounding celestial sphere, which is all around us and whose radius for all practical purposes is taken to be infinitely large**, say 2,000 light-years or more. Distances to the stars/galaxies are another matter. The visible stars to the naked eye are at distances from 4 light-years to approximately 2,000 light-years. By just looking at a star, one cannot tell what the distance is to that star.

It is OK for everyone to consider that she/he is at the center of the **celestial sphere** because **the radius of the celestial sphere is infinitely long** when compared to the size of even the whole solar system. Therefore, everything on Earth and even on the total Earth's orbit around 'our' Sun (orbital radius is 500 light seconds = Astronomical Unit) can be considered to be practically one and the same tiny point at the center of the infinitely large surrounding celestial sphere.

Now, don't feel small, because we all are very important! **For comparison, as was said, all naked-eye stars are within** approximately **a 2,000 light-year distance. This 2,000 light-year long distance is approximately 1.6 million times longer than the diameter of the orbit of planet Pluto.** Therefore, the whole Solar system can be considered to be just a point when compared even to the 2,000 light-year distance. Using telescopes, the most distant objects seen (2007) in the universe are almost 13.7 billion light-years away from us.

Google: < celestial coordinates >
                 < celestial declinations and right ascensions >
                 < astronomical distances >

7. Star Constellations

Ancient astronomers selected and named many star constellations. They grouped a number of seemingly nearby stars into small 'areas' covering most of the surface of the celestial sphere. These areas in the sky are called constellations. The sky (celestial sphere) is covered by approximately one hundred constellations. A soccer ball covered by pentagon and hexagon shaped surface areas provides a good mental image of the constellations on the celestial sphere.

Animal names for these constellation areas are common. Individual stars in one constellation may be anywhere from 4.2 to 2,000 light-years away from us. The ancient astronomers did not know that. The brightest star in a constellation often has a prefix alpha, the second brightest star often has a prefix beta, and so on.

Constellations are used nowadays to give a general direction to an object in the sky that can be at any distance up to approximately **14 billion light-years away from us**.

Horoscopes use twelve constellations, which are in a band near the ecliptic plane = Earth's extended orbital plane onto the skies all the way to the celestial sphere. The radius (radius vector) from the Sun to Earth sweeps that plane in one year.

For instance, the North Star, or Polaris, is in the constellation Ursa Minor (Small Bear), and the Big Dipper is in Ursa Major (Big Bear) constellation. Instead of the imprecise general directions to vague constellations, astronomers use more precise celestial coordinates, which are somewhat similar to latitudes and longitudes on Earth. More precise celestial coordinates are needed to get a desired celestial object in the crosshairs of a telescope. The names of constellations are used only to indicate the approximate general area in the sky where an object is seen.

Google: < alphabetical listing of star constellations >
                 < polaris or north star or pole star >
                 < celestial coordinate system >
                 < ecliptic constellations >
                 < the origin of the zodiac >

8. THE SUN'S LOCATION IN THE MILKY WAY GALAXY

We, and all visible stars (suns) to the naked eye, are in the same pinwheel arm of the Milky Way Galaxy. The nearest visible naked-eye star is Proxima Centauri (in constellation Centaurus) at a distance of 4.24 light-years = 275,000 times as far as 'our' Sun is from Earth. Beyond the 2,000 light-year naked-eye distance limit, the only visible celestial object (with positive declination) is a hazy patch of the neighboring disk-shaped galaxy Andromeda.

Google: **< the 50 nearest stars >** Scroll down to the table.

< star magnitudes > Star magnitude describes star's apparent brightness.

< milky way galaxy >

< news about the milky way galaxy >

To get a good grasp of our location in this Milky Way Galaxy, **consider that a 12-inch (30 centimeter) diameter pizza, Frisbee, a one-inch thick round sheet of foam or a piece of cardboard represents the disk of the Milky Way Galaxy.** Cut it so that pinwheel arms and its 3-inch center part remain. Further consider that a one half-inch (12-millimeter) diameter whole is cut/drilled/punched in the pizza/Frisbee/foam/cardboard disk about three inches (75 millimeters) from its edge into one of the arms. Put a half-inch diameter marble into the hole. This marble represents the 4,000-light-year **diameter** celestial sphere, where all the visible naked-eye stars are, and Earth and all solar planets 'together' with 'our' Sun are at the tiny center point of that marble. So, there we are in this Milky Way Galaxy!

As was mentioned, all the couple of thousand 'naked-eye' stars are within the marble placed in the hole. The diameter of that hole is 4,000 light-years or approximately 4% of the 100,000 light-year diameter of the Milky Way Galaxy disk. The total number of stars in the Milky Way Galaxy is estimated to be between 100 billion and 400 billion (= 100 x 10E9 to 400 x 10E9).

There are many more than 'those 2,000' naked-eye stars in that marble-size hole in the Milky Way model because many stars are not bright enough to be seen without telescopes. The mutual distances between naked-eye stars in one constellation can be any distance up to 2,000 light-years. When all directions (constellations) are considered, distances between two visible stars can be up to 4,000 light-years, if they are in opposite directions, each 2,000 light-years from 'here'.

The total area of the Milky Way Galaxy disk is approximately 625 times greater than the area of the little 4,000 light-year diameter marble hole (2 x 2,000 = 4,000) depicted in the pizza/foam/Frisbee/cardboard model. The Milky Way is thicker near its center than it is in 'our' pinwheel arm.

The largest known star 'KY Cygni' has a diameter 1300 times the diameter of 'our' Sun and larger than the orbit of Jupiter.

Google: **< the milky way galaxy >**
              < seti >
              < astronomers discover largest stars known >

8.1. PLANET EARTH IN THE UNIVERSE

Earth and this whole solar system are in one pinwheel arm of the Milky Way Galaxy.

Google: **< the milky way galaxy – our home >**

Together with 'our' Sun and the 'near-by' stars in this pinwheel arm, we are orbiting (going around) the Milky Way galactic center at a linear speed of approximately 250 kilometers = 160 miles per second.

In addition to its spinning, the whole Milky Way Galaxy is on the move with respect to other galaxies, and in turn, they are also moving – 'God knows where'. Is it wobbling as it goes? Nobody knows. The linear speed of the Milky Way depends on which other galaxy Milky Way's speed is compared/referred. The range of those mutual speeds of galaxies can be from zero to 100 kilometers = 60 miles per second to speeds approaching the velocity of light.

Planet Earth is 'tied' to 'our' Sun by its gravitational pull (attraction). Earth has kept and will keep its distance to 'our' Sun like being on Sun's leash at a distance of one Astronomical Unit for eons. Earth's centrifugal acceleration balances Sun's gravitational pull as a result of Earth's orbital speed of approximately 30 kilometers (18.6 miles) per second. If Earth's orbital speed would be brought to a standstill, Earth would dive into 'our' Sun in a couple of months.

Recall from earlier that the mutual distance *between Andromeda and Milky Way galaxies* is decreasing by approximately 300 kilometers (200 miles) per second, but because we are situated in one rotating Milky Way pinwheel-arm, we are (now) moving in the general direction away from the direction to Andromeda at approximately some 200 kilometers (100 miles) per second, and **the relative speed** (as measured from Earth) to Andromeda is the difference, or about 100 kilometers (60 miles) per second.

After approximately another 120 million Earth-years = after one-half galactic year, the measured (from Earth) approaching speed of Andromeda will be 300 + 200 kilometers (200 + 100 miles) per second, or approximately = 500 kilometers (300 miles) per second. The mutual distance between these two galaxies will still be decreasing by approximately 300 kilometers = (220 miles) per second also at that time.

Still another 'thing' must be considered. This Earth also orbits 'our' Sun at 30 kilometers (20 miles) per second, and sometime during one year, Earth moves in the general direction toward Andromeda and then six months later in the opposite direction. For this reason the measured relative approaching speed between this solar system and Andromeda will vary by plus-minus something a little less than the 30 kilometers (20 miles) per second. Even this must be considered to get the mentioned average approach speed of 100 kilometers (60 miles) between these two galaxies.

There is still more to these approach velocities. Andromeda is not in Earth's orbital plane because its declination is about +41 degrees with its Right Ascension of about 0 hours 43 minutes. For this reason, the 'from-Earth-measured approach speed' is also depending on these directional differences. **These**

details will not be dealt with in this book. This general information is beneficial for understanding 'the whole big picture' of Earth and its motions in the neighboring universe.

Google: < celestial coordinates >

               < celestial declinations and right ascensions >

We all are truly fortunate and lucky to be living on this unique wonderful planet Earth in this solar system. It would be almost impossible to even imagine a better place for Earth's location. According to some estimates, humans have been living here for 150,000 years but Earth itself is much older. **Earth's age is estimated to be 4,550,000,000 (4.55 billion) years. The universe has been estimated to be** approximately **eleven to twenty billion years old with 13.7 billion being the most recent (2007) estimate.**

Google: < age of the universe >

               < HubbleSite-faqs.how old is the solar system? >

               < redshift >

               < hot big bang >

http://chandra.harvard.edu/photo/chronological. html

9. Possible Alien Visits to Earth

This chapter describes why there cannot be any visits by any living Aliens to this Earth. Alien (extra terrestrial) human-type life is possible on some planets orbiting some 'unreachable stars' (suns) located at distances at least 8 light-years away from us. Radio/laser communications with some Milky Way Aliens may be possible but it will be painstakingly slow since it takes more than 8 years for the message to travel each way. All one-way spaceship travels would take centuries even at speeds of 3,000 kilometers (2,000 miles) per second, which is only approximately 1 % of the speed of light. The required energies to propel a spaceship to a speed of 3,000 kilometers (2.000 miles) per second, or more, are not available.

All stars from here to at least 8 light-year distances have been classified as unfriendly for having life-supporting planets. Therefore, other solar systems with life-supporting planets must be at greater distances than the 8 light-years.

Search for Extra Terrestrial Intelligence (SETI) is an ongoing project. Hopefully, contacts will succeed in the future with the residents of those 'far away places'.

Why the 8 light-year limit? There are **mainly two reasons** for it.

First, the star with orbiting planets, on which life is possible, must be a single star and not a doublet (two stars orbiting around their common center of gravity (gravitation). Approximately **10 % of Milky Way 100 billion stars are doublets.** If a planet is close enough to a doublet to be warm enough to allow the existence of liquid water, the planet will eventually crash into one of the doublets because the **'Three Body System'** is an unstable system. Planets orbiting farther away at survival distances from the doublet are too cold to have liquid water, even if they happen to have water molecules.

44

Secondly, the planet's star (sun) should radiate just about the right kind and intensity of energy. All single stars are not similar to 'our' life supporting Sun. The Sun's radiation should not be too strong nor too weak. In other words, the star should be about a Sun-like, 'Goldilocks-type' star.

Google: < three body problem is celestial mechanics >

               < double stars >

               < the double star library >

               < messier object 31 > = Andromeda, the nearest large spiral galaxy.

Other galaxies in the known universe are at distances up to 13.7 billion light-years away, and therefore all other galaxies are completely out of reach for even any type of two-way radio/laser contact. Only a very small portion of this Milky Way Galaxy is the only 'small' volume in the whole universe where Extra Terrestrial life can even be studied. Remember that a one-way radio/laser signal to/from other galaxies takes millions of years.

10. Common Velocities in Space

All bodies (masses) in space are moving all the time in their own directions. The driving forces originate from gravitational pulls and radiation pressures from surrounding objects. If there would be a stationary object somewhere in space, it would start moving due to some gravitational attractions (accelerations)/radiation pressures that are present everywhere in the universe.

This book describes **the high speed this Earth has in the vacuum of space in its annual orbit around 'our' Sun. We are traveling at 20 miles per second = 30 kilometers per second all the time.** Only a few people seem to know/realize that we all are constantly 'flying' that fast through the surrounding space.

Kepler's Laws describe the planetary orbits/velocities/periods. Of the nine solar planets, Mercury (nearest to the Sun) moves the fastest, Pluto (farthest from the Sun) moves the slowest.

The average orbital velocities of solar planets are:

Mercury	47.9 km/s	= 29.8 miles/second
Venus	35.0 km/s	= 21.7 miles/second
Earth	29.8 km/s	= 18.5 miles/second
Mars	24.1 km/s	= 15.0 miles/second
Jupiter	13.1 km/s	= 8.1 miles/second
Saturn	9.6 km/s	= 6.0 miles/second
Uranus	6.8 km/s	= 4.2 miles/second
Neptune	5.4 km/s	= 3.4 miles/second
Pluto	4.7 km/s	= 2.9 miles/second

The average speed of 'our' Moon in its monthly orbit around this Earth is approximately one kilometer (0.6 miles) per second.

Solar particles (protons, etc) from the Sun are usually traveling at speeds of a few hundred kilometers (miles) per second. Common speeds are between 200 to 700 kilometers per second = 125 to 450 miles per second. A very high speed of 2,235 kilometers = 1,389 miles per second = almost 0.1 % of the speed of light was recorded in November 2003.

Google: < solar wind velocity >

In this Milky Way Galaxy 'our' solar system is orbiting the Milky Way center at a speed of approximately 250 kilometers (160 miles) per second going once around in 225 million years, which time period is called one galactic year.

Many stars near to the center of Milky Way Galaxy have orbital speeds in excess of 5,000 kilometers (3,000 miles) per second. There may be asteroids traveling between the stars at high velocities covering a distance equal to one Astronomical Unit (= 149,600,000 km = 93,000,000 miles = Earth-to-Sun distance) in less than ten hours. They are very few and far between in our neighborhood. Comet McNaught, (perihelion on January 12, 2007), was fast enough to never return.

http://www.spaceweather.com/comets/gallery_mcnaught.php

Some distances between many galaxies are changing almost at the speed of light 299,792.458 kilometers (186,282.025 miles) per second. Velocity of light is the absolute maximum speed limit for all material objects.

Most velocities and distances to moons, planets, stars and galaxies are obtained by Doppler Shift measurements of the incoming radiation (light, infrared, X-rays) from an object due to the fact that the wavelength of the emitted radiation by an approaching, or receding moving object is shifted.

The incoming radiation is blue shifted, if the distance to the object is decreasing and the incoming light (radiation) is shifted toward shorter wavelengths.

The incoming radiation is red shifted, if the distance to the object is increasing and the incoming light (radiation) is shifted toward longer wavelengths.

Google: < solar system data >

            < Primer on the Solar Space Environment >

            < earth and moon viewer >

            < earth and sun viewer >

            < solar system viewer > Click on Inner Solar system, then on Asteroid belt, then on Outer Solar system, then on Kuiper Belt and then on Comets.

            < galactic viewer >

            < number of galaxies in the universe >

            < Doppler shift >

            < spectral lines in stars >

11. EARTH AS A SPHERE

E arth's surface is not flat, except for some small areas here and there. The flat areas do not extend very far. Topographic elevation and ocean depth features are relatively small when compared to the whole Earth.

The average (mean) radius of Earth is 6371 kilometers = 3959 miles.

The volume of Earth is 1.08 x E12 cubic kilometers or 2.60 x E11 cubic miles.

The area of Earth's surface is 510 million square kilometers = 197 million square miles = 5.1 x 10E10 hectares = 1.04 x E12 acres including the oceans.

The land areas are 29 % of the total, or 3.02 x E11 acres = 1.48 x 10E10 hectares. If the Earth's population is 6 billion, the land-area per person is approximately 50 acres = 2.5 hectares = 25,000 square meters.

The highest mountaintop, Mount Everest, is 8,848 meters (29,028 feet) above the mean (average) sea level. It is 0.14% of Earth's mean radius. On a 1:5,000,000 scale model of Earth, Mount Everest would only be a 1.8-millimeter (0.07-inch) high ripple.

The deepest depth in all of Earth's oceans is in Mariana Trench in the Pacific Ocean. The deepest measured water depth there is approximately **11,035 meters (36,204 feet) or 0.17% of the mean radius of Earth**. On a 1:5,000,000 scale model of Earth, this deepest spot in the ocean would be only a 2.2-millimeter (0.09-inch) deep dent. The water pressure at the deepest ocean floor is approximately 1,100 times the atmospheric pressure at the sea level.

Earth is bulging near its Equator, i.e. the equatorial diameter is longer than the polar diameter. Because Earth is not a perfectly solid, unyielding rigid body, it is flattened due to the centrifugal acceleration of Earth's daily (diurnal) rotation (spinning) around its north-south axis.

The polar radius (the distance from the equatorial plane to the poles at the Ellipsoid level) of the Earth Ellipsoid is **6,356.752 kilometers = 3,949.895**

miles. It differs from Earth's mean radius (6371 kilometers = 3959 miles) by 0.22 %, and the **equatorial radius** of the Earth Ellipsoid is **6,378.140 kilometers = 3963 miles** differing from Earth's mean radius by 0.11 %.

Google: < the earth as an ellipsoid >

For many practical purposes, when discussing these dimensions with a general 99% percent precision, as for instance in this book, **Earth can be considered to be a sphere with an average (mean) radius of 6371 kilometers = 3959 miles.** Millimeter accuracies and the small meridian flattening are sometimes important, but not for the generalities described in this book.

That part of Earth Sciences, which studies the details of **accurate positioning of points** (azimuths from point to point, latitude, longitude, elevation, or XYZ-coordinates) on and near Earth's surface **and tries to determine Earth's exact size and detailed shape, is called Geodesy**, or Geodetic Science.

Among the positioning tasks of various (map) points on Earth's surface, **there are two main types of geometrical problems.** This book does not discuss the precision ellipsoidal trigonometry and the mathematics needed to solve these problems beyond mentioning their existence.

11.1. THE DIRECT PROBLEM IN COMPUTING EARTHLY DISTANCES

For instance, from a starting point (assume its coordinates are known on Earth's surface) walk 100 yards (meters) in NE-compass direction, which is in 45-degree azimuth. The task is to determine the coordinates of the end point. After arriving at the end point, 'turn 180' and retrace your steps to the starting point. The back-azimuth from the end point is approximately 225 degrees or in SW-compass direction in which the retracing proceeds. The Direct Problem is solved after the coordinates of the end point and the back-azimuth are computed.

Great Circle distances can be used over 'short' distances and for approximate numerical values, but not for all 'government' work. A Great Circle is the circle cut by a plane running through the center of the sphere (Earth).

Spherical Trigonometry is used to solve the Great Circle problems. Accurate work over 'long' distances requires Ellipsoidal Trigonometry on the surface of the Earth Ellipsoid.

Generally for longer distances: starting from a known point in a given direction (azimuth) over a given distance, one has to determine the coordinates of the end point of this geodetic line (shortest possible distance along Earth's surface) and also the back-azimuth. The back-azimuth is the direction (azimuth) of the geodetic line from the end point back to the initial point. After the geodetic (geographic) coordinates of the end point and the back-azimuth from the end point have been computed, the Direct Problem is solved.

Google: < **Geodetic Survey Division – Products – Software – INDIR** >
               < **geodetic direct problem** >
               < **back azimuth** >
               < **datum and earth ellipsoid** >
               < **geodetic line** >

11.2. THE INVERSE PROBLEM IN COMPUTING EARTHLY DISTANCES

The solution answers the questions: 'What is the distance along Earth's surface between two known (given) points, and what are the azimuths (bearings, directions) from each of these points to the other point?'

The latitudes, longitudes and topographic elevations of two points on Earth's surface are given. For the Direct as well as the Inverse Problem these values must be reduced to sea level for both ends of the line and then to the chosen Earth Ellipsoid, along which mathematical computations can be done. After the ellipsoidal distance between the two known points, as well as the azimuth and the back-azimuth have been computed, the Inverse Problem is solved. The sea level reductions are a much smaller matter.

One example for instance, of the Inverse Problem is flying from Chicago, Illinois to Rome, Italy. What is the distance to Rome, and what is the shortest possible distance from Chicago to Rome? Distances between airport runways are not needed to millimeter accuracies, but in trying to measure continental wanderings, millimeter accuracies are needed.

Another example of the Inverse Problem is to find out if the 'given' (assumed) azimuth (direction) to Mecca, Saudi Arabia from Chicago, Illinois is correct. If the direction is wrong, one may be facing Rome, Tel Aviv, or the North Pole.

These two main problems of Geometric Geodesy are an important part in providing the framework for accurate mapping and navigation. Many great mathematicians (Bessel, Helmert, Gauss, Clairaut, Laplace, Lagrange, Legendre, and others) devoted some of their time and effort trying to find solutions to these two problems for the surface of an ellipsoid of revolution.

One can wonder what those wise men would have come up with had they had today's computers, GPS and satellites at their disposal. Electronic computers for this type of tasks became commercially available for general use around 1970. How lucky we are compared to the mentioned famous mathematicians! They deserve a tip of the hat together with some earlier Earth Scientists listed below as Google-addresses.

Google: < geodesy >
               < international association of geodesy >
               < important geoscientists >
               < friedrich robert helmert > 1843-1917.
               < friedrich wilhelm bessel > 1784-1846.
               < karl friedrich gauss > 1777-1855.
             . < adrien-marie legendre > 1752-1833
               < pierre-simon laplace > 1749-1827.
               < joseph-louis lagrange > 1736-1813.
               < alexis claude clairaut > 1713-1765.
               < johannes kepler > 1571-1630
               < tycho brahe > 1546-1601
               < nicolaus Copernicus > 1473-1543.
               < erathostenes' method > ?-194

11.3. GENERAL SHAPE OF THE EARTH

Only a few of us keep in mind in our daily lives the fact that this planet Earth is only a small spinning round globe flying through the emptiness of space in its annual journey (orbit) around 'our' Sun at an **average orbital speed of 30 kilometers (18.6 miles) per second or 107,230 kilometers = 66,630 miles per hour.** The Sun's steady gravitational pull keeps and has kept Earth in its livable orbit year after year for a few billion years.

Although the **deviations in the shape of Earth from a perfect sphere are small**, they are important features to understand for several reasons. **This general deviation from a perfect sphere is expressed by the flattening, or by the eccentricity of the meridian ellipse, or by the equatorial bulge.**

All ellipsoidal meridian ellipses on a rotational ellipsoid have the same size, flattening and eccentricity. When any meridian ellipse is rotated around Earth's spin axis, it traces a surface called **the Earth Ellipsoid. It is also called Ellipsoid of Revolution, although the more correct name is the Rotation Earth Ellipsoid. Rotation Earth Ellipsoid has a definite mathematical shape and an easy mathematical formula describing its surface usually in XYZ-coordinates.**

The actual mathematical Earth Ellipsoid surface is chosen to approximate the mean sea level surface (= Geoid) as well as possible. The mean sea level surface (=Geoid) undulates over and under the mathematical Earth Ellipsoid by amounts, which are usually less than 100 meters (300 feet), and they are called Geoid Undulations.

Google:　　< geoid map >

The word Spheroid is sometimes used to describe the shape of the body of Earth. The word spheroid actually means an unspecified vague shape approximating a sphere without a definite given mathematical formula.

Google:　　< world geodetic system [wgs] >

11.4. ROOM-SIZE MINIATURE BALLOON MODEL OF THIS EARTH

The following model shows how very close the mathematical Earth Ellipsoid is to a perfect sphere.

A good visual image, and an appreciation and understanding of the amount of Earth's flattening and some other terrestrial features, can be obtained by considering the following **miniature model** of Earth.

This three-dimensional model depicts Earth in a scale of one to five million. A one to five million-scale map of the continental United States is quite common. On a map of that scale, a distance of 5,000 kilometers = five million meters (3106 miles) is one meter = 3.28 feet = 39.37 inches, which is the approximate coast-to-coast map distance over the 48 conterminous states of the United States of America.

If any actual **linear distance on this Miniature Model Globe** is multiplied by five million, the actual corresponding distance is obtained on Earth's surface.

The areas on this model globe are smaller by five million squared (= 2.5 x 10E13) when compared to the same actual areas on Earth.

The volumes are five million cubed (= 1.25 x 10E20) times smaller than the corresponding volumes on the actual Earth.

Consider a perfectly spherical balloon having a diameter of one five millionth of Earth's mean diameter, or 2 x 6,371,200 meters divided by 5,000,000 = 2.548 meters = 254.8 centimeters = 8.361 feet.

Consider this perfectly spherical balloon to be placed in the middle of a large room with a 10 foot = 3.048 meter = 304.8 centimeters high ceiling. Vertically, this balloon touches the floor, and its highest part leaves exactly 304.8 − 254.8 = 50-centimeter (19.7 inches) space to the ceiling.

Next shorten the polar diameter so that the flattening of the balloon will be the same as Earth's actual flattening, which gravity might do almost automatically for the balloon. The result is that the polar diameter will have to be 8.54 millimeters (0.336 inches) shorter than the equatorial diameter. The new polar diameter of the slightly flattened balloon will be 2.53945 meters and the equatorial diameter will remain at 2.548 meters. The flattening of the deformed (flattened) balloon is obtained by dividing the difference between the equatorial and polar radii (2.548 minus 2.53945) by the equatorial radius 2.548, which results in 0.00335281. For comparison, the World Geodetic system 1984 gives the flattening as 0.00335281. **Thus, the chosen model is a very good approximation for the actual Earth Ellipsoid.**

Google: **< wgs 1984 >**

By looking at this slightly flattened balloon in the big room, it would be very difficult to tell whether the balloon is flattened by 8.54 millimeters (0.336 inches) or not.

Earth's atmosphere is just a very thin shell (layer) surrounding the ground and ocean surfaces. Approximately **one half of the atmospheric mass is under an altitude of** approximately **5.6 kilometers = 3.5 miles = 18,000 feet corresponding to 1.1 millimeters (0.044 inches) thick (thin?) layer of spherical shell around the room-size balloon model. Earth's outer atmosphere reaches to over 560 kilometers = 350 miles from Earth's surface or to** approximately **11 millimeters = 0.44 inches for the model.** Further out, there is the vacuum of space with very few air molecules.

11.5. EARTH'S ROOM-SIZE BALLOON MODEL AND THE ACTUAL EARTH

The one to 5 million scale model shows that this Earth is a very smooth round planet when seen from outer space.

NASA Shuttles have been orbiting Earth at altitudes between 190 and 360 miles = 300 and 580 kilometers.

Dividing these altitudes by five million, one gets 6 to 11 centimeters = 2.4 to 4.3 inches for the Shuttle orbiting altitudes for the room-size balloon model. This means that if one walks so close to the balloon that the eyeballs are 2 to 4 inches from the balloon, the balloon is at the similar relative distance as the Shuttle astronauts have seen this Earth from their orbits.

Because the normal human eye does not see objects clearly within four inches from the eye, consider looking at the balloon surface from 10 inches = 25 centimeters ahead to see Earth as the Shuttle astronauts have seen it.

Even at the Shuttle Orbiter altitudes, there are still many air molecules caus-ing a small air drag slightly reducing the velocities of the Shuttles. This velocity reduction (deceleration) from the air-braking action causes a weak micro-grav-ity for/in the Shuttle. If nothing would be done to keep it in its desired orbit, it would finally cork-screw itself down into the denser lower atmosphere, and would finally come down due to this air-braking and to many other perturba-tions affecting its orbit. Of course, all Shuttle landings from their orbits have been controlled landings except one. (The Challenger did not reach its orbit).

Orbit maintenance (pushing them back to their desired orbits) is also re-quired to keep most functioning satellites in their useful orbits, including the communication/weather and the GPS (Global Positioning System) satellites.

GPS satellites orbit in almost circular orbits at altitudes of about 10,900 miles = 17,540 kilometers above Earth's surface or at distances of about 22,200 to 23,900 kilometers = 13,800 to 14,850 miles from Earth's center. Various sources may give slightly different numbers.

The communication satellites are in 24-hour (sidereal hours) almost circular, al-most equatorial orbits at altitudes of approximately 22,223 miles = 35,764 kilometers or at approximately 42,136 kilometers = 26,182 miles from Earth's center.

Google: **< orbital perturbations of satellite orbits >**
                < US Naval Observatory (USNO) GPS Operations >
                < NGS-PRECISE GPS ORBITS >
                < location of GEO communication satellites >

11.6. ROOM-SIZE BALLOON MODEL AND THE MOON ASTRONAUTS

When the Moon astronauts were traveling at an altitude of 22,223 miles = 35,764 kilometers above Earth's surface, (or from Earth's center at a distance of 42,227 kilometers = 26,239 miles, or approximately = 6.628 times Earth's radius), where the 24 hour sidereal communication satellites orbit, the apparent size of Earth they saw was the same as the room-size balloon model is seen from a distance of 7.1 meters = 23.4 feet. The angular size of Earth's diameter from that distance is approximately 17 degrees.

The visual size of Earth from there was approximately the same as looking at the mentioned 8.4 foot (2.5 meter) diameter balloon model from a distance of 28 feet (8.4 meters) from the center of the balloon or from 24 feet (7 meters) from the balloon's nearest surface. This Earth looked pretty big to them, didn't it!

When the astronauts were on the Moon's surface, the apparent size of Earth was the same as one sees the room-size balloon model from a distance of 76 meters = 250 feet. The angular size of Earth's diameter as seen from the Moon is 1.9 degrees, or Earth seems to be 3.7 times as large as the Moon is seen from Earth.

See the Earth from the Moon at Internet address:
< www.space.com/imageoftheday/image_of_day_041004.html >

A good mental image of these angles (1.9 and 17.4 degrees) is obtained from a tilted shelf on a wall. If one end of a 60 inches = 5 feet = 152 centimeters long horizontal shelf is lifted up by 2 inches = 5 centimeters, then the shelf is tilted by 1.9 degrees. If one end of the shelf is tilted up by 18.8 inches = 48 centimeters, the shelf is tilted by 17.4 degrees.

From the Sun's distance, or from one AU = 1 AU = 149,600,000 kilometers = 92,957 miles, Earth itself would look like the room-size balloon model looks from a distance of 18.6 miles = 30 kilometers. The angular size of Earth's diameter from 1 AU distance is 8.8 seconds of arc, or the same as a dime ($0.10) is seen from a distance of 963 feet = 294 meters.

12. Miniature Models of the Entire Universe

The following is a real, actual and easily-visualized three-dimensional approximate model of the whole known universe. It is a **previously unpublished miniature model** (as far as the author knows) with all estimated galaxies included. The first model is a conglomerate cube made of one cubic foot (12 inches x 12 inches x 12 inches) cardboard boxes stacked to form a big conglomerate cube having sideline edges 4.27 miles = 22,568 feet = 6879 meters long. We could walk around it, fly over it in an airplane and look at that miniature three-dimensional model of the whole known universe from its outside.

The model is described in three versions; the smallest one would fit in many backyards. **Its sideline edges of the conglomerate cube are only 74 feet = 22.6 meters long.**

These models make it possible to look at the whole known universe from its outside! Very few have ever before seen anything so big in such a small volume.

The described models will have the same total scale volume as the previously estimated total spherical volume of the total known universe (10E31 cubic light-years), but it will have a cubical shape to make 'things' easy and reasonable to describe and understand. The model will be a good and simple model.

There might be a temptation to think that this is the way God sees it, but of course, we cannot know, nor correctly guess, what our Creator/God sees.

12.1. Miniature Model Number One of the Known Universe

To deal with full inches and feet in the model, the average distance between adjacent galaxies is taken to be at least one million (10E6) light-years, or twelve inches in the model, although it may be 1.2 million light-years plus/minus some.

This still will produce a good 'ballpark' model. Even the best estimates of this type of numbers for the universe (average number of galaxies per cubic light-year) are just estimates based on available samples and they are subject to some revisions as more observations are made in the future.

As mentioned, the following is a real, actual and easily-visualized three-dimensional approximate model of the known universe. It is a conglomerate cube made of one cubic foot (12 inches x 12 inches x 12 inches) cardboard boxes stacked to form a big conglomerate cube having sideline edges 4.27 miles = 22,568 feet = 6879 meters long.

Two smaller similar models (Model Number Two and Model Number Three) will also be described.

The sideline edges of the Model Number Two conglomerate cube are only 1,881 feet = 573 meters long.

The sideline edges of the Model Number Three conglomerate cube are only 74 feet = 22.6 meters.

To deal with full inches and feet in the model, the average distance between adjacent galaxies is first taken to be one million (10E6) light-years, or twelve inches in the model, although it may be 1.2 million light-years, or even up to five light-years in some volumes of the universe. **This will still produce a good 'ballpark' model.** Even the best estimates of this type of numbers for the universe (average number of galaxies per cubic light-year) are just estimates based on available samples, and there will be revisions as more observations are made in the future. Internet site 'http://www.answers.com/topic/list-of-nearest-galaxies' lists 39 nearest galaxies within 3 light-year distance, or 39 in 118 cubic light-years, or one galaxy in every 3 cubic light-years.

Recall that **the volume of the known (spherical) universe** with its radius of approximately 14-billion **light-years is (1.15 x 10E31) cubic light-years.** The conglomerate cube will have this same scale volume.

1. The one-foot long sideline of each cubic foot box represents a million light-years. The **volume of each box** represents 1,000,000 x 1,000,000 x 1,000,000 = **10E18 cubic light-years.** Therefore, the total volume of the universe = 1.15 x 10E31 divided by 10E18 is = **1.15 x 10E13 is the total number of required cubic foot cardboard boxes** for

this model. The model will have them. This model is generously big enough to have up to 90 boxes for each galaxy, if the total number of galaxies in the universe is 125 billion.

2. Assume a quarter coin ($0.25) is inside (near the center) of each cubic foot box. Each quarter coin with its diameter of almost one inch represents one average galaxy with a diameter of 100,000 (10E5) light-years, such as 'our' Milky Way Galaxy. Then the distance between neighboring galaxies will be approximately one million light-years.

3. Consider that these 1.15 x 10E13 boxes are piled next to each other horizontally and vertically without empty spaces in between them to form a large conglomerate cube, which – as it will turn out – will have its sideline edges 22,568 feet = 4.27 miles = 6.9 kilometers long.

Andromeda Galaxy, the nearest large spiral disk galaxy, is at a distance of approximately three million light-years (three cubic foot cardboard boxes) away from us, i.e. from the Milky Way Galaxy. There are also galaxies, which are in the process of colliding with each other, practically occupying the same volume.

To find the length of the sideline of the large conglomerate cube, which has the same volume as these (1.15 x 10E13) cardboard boxes (= the total volume of the known universe), one has to take the cubic root of (1.15 x 10E13). This can be done with a pocket calculator. The result is that the sideline edges of the large conglomerate cubic box are 22,570 feet long. To check, raise this 22,570 to third power (= 22,570 x 22,570 x 22,570), and get (1.15 x 10E13), as you should.

Because one mile is 5280 feet, the 22,570 feet is 4.27 miles = 6.9 kilometers. **The 4.27 miles = 6.9 kilometers is the width, length and height of the conglomerate Miniature Model Number One.** It has the same representative volume as the universe. Each box could contain one or more galaxies, or have no galaxies in it.

Because there are 1.15 x 10E13 boxes in the model and if the estimated number of galaxies in the universe is 125 billion, there are 92 boxes for each galaxy in the model. The total number of galaxies could be larger than the 125 billion. Anyway, the volume of the model was selected have the same volume as the universe has. Some clusters of galaxies may have thousands of galaxies in them.

Voila, here we have a real three-dimensional approximate model of the whole known universe! We all can easily imagine a cube with its sidelines of 4.27 miles (6.9 kilometers) long. For economical and other reasons, it is not practical to actually build this model, but the size and shape can be easily visualized in our minds. The cardboard boxes are not free, and about 125 billion or more quarter coins would be needed. These quarter coins alone would be worth 32 billion dollars, if every box gets a quarter. Outside Washington, D.C. that is an astronomical number! No one could/would spend money for the actual model, but the boxes and the quarter coins are helpful for the description of the model.

The *linear map scale* of this Model Number One is one to 3 x 10E22. This is obtained as follows:

The one-foot sideline is 12 inches = 0.3048 meters and it represents one million light years.

One light-year is 9.46 x 10E15 meters and therefore, one million light-years are 9.46 x10E21 meters.

The scale is then as 0.3048 meters is to 9.46 x 10E21 meters, or one to 2.9 x 10E21, which rounds to scale: one to 3 x 10E22.

Google: < one light year >

12.2. LOCATION OF PLANET EARTH IN THIS MODEL

The quarter coin representing the Milky Way Galaxy in this model is in one cubic-foot cardboard boxes near the center of the large conglomerate cube at a distance of some two miles (plus/minus) from each outside face of the large conglomerate cube. If the Milky Way Galaxy is not that near to the center of the conglomerate cube, it is simply farther from the center, which is OK for the model.

In the model 'our' Solar System, i.e. 'our' Sun itself with planet Earth nearby is near George Washington's upper forehead of the quarter coin representing 'our' Milky Way Galaxy. Remember that the diameter of the quarter coin represents 100,000 light-years, and that the Earth is only a distance of 500 light-seconds from 'our' Sun.

12.3. MINIATURE MODEL NUMBER TWO OF THE UNIVERSE

If one-inch cube cardboard boxes had been selected instead of the one cubic foot boxes, the conglomerate large cube would still contain the same number of the smaller cubic inch boxes, but **all linear distances would shrink by a factor of 12**, and then the **total length of the sidelines of the large conglomerate cube would be only 22,570 feet divided by 12, or = 1,880 feet = 573 meters = 0.36 miles = 0.57 kilometers.**

In this Model Number Two having the **one cubic inch boxes,** the diameters of the quarter coins must also shrink to 2 millimeters, or to the size of a baby bedbug, which may be substituted for the quarter coins.

Google: < bedbugs >

The cubic conglomerate model (1880 feet by 1880 feet by 1880 feet) made up of one cubic inch boxes is not much taller than the Taipei Financial Center in Taipei, Taiwan (1667 feet = 508 meters) or the Sears Tower in Chicago, Illinois (1450 feet = 442 meters).

The linear map scale of the second model, where one inch represents one million light years, is one to 4 x 10E23.

The two models of the universe described above **have the shape of a cube. Nobody knows whether the actual shape of the universe is spherical, ellipsoidal, cubic, potato-like or any other shape.** How about an octopus?

12.4. MINIATURE OVERVIEW MODEL NUMBER THREE OF THE UNIVERSE

Had one-millimeter cubical paper boxes been selected for one cubic light-year volume instead of the one cubic inch boxes, the conglomerate large cube would still contain the same number of the smaller size cubic millimeter boxes representing the total volume of the known universe. For one cubic millimeter boxes, **all linear distances would shrink by a factor of 25.4** (one inch = 25.4 millimeters), and then the **total length of the sidelines of the large conglomerate cube of the whole known universe would be 1880 feet divided by 25.4, or**

only 74 feet = 22 meters. It is literally *The Real Overview Model*.

For a numerical check, by raising this 22,600 to power three, one obtains 1.15 x 10E13, which is the total number of cubic boxes in these three models and also the number of one cubic millimeter paper boxes in the 'backyard' model of the known universe.

The linear scale of this model is one to 1.0 x 10E25.

The three described and previously unpublished cubical models depict the universe in a manner, which can be imagined and understood much better than not having a model at all. **The cubical models are reasonable approximations for the size and shape of the universe.**

12.5. WHERE ARE WE IN THIS UNIVERSE?

The presently-known universe extends from here (Earth) in all directions (think of all the radii emanating from the center of a spherical ball: basketball, etc.) to 14 billion (14,000,000,000) light-years. The latest (2007) number seems to be 13.7 billion light-years. Whether our home galaxy, the Milky Way, is near (or how close) to the center of the known universe is not known for sure, but here we all are.

Recall that one light-year is the distance electromagnetic radiation (light) travels through vacuum in one year, or 1 light-year = 9.46 x 10E12 kilometers = 5.88 x 10E12 miles = 63,000 AU (Astronomical Units).

The radius of the known universe is 13.7 billion light-years = 1.3 x 10E23 kilometers = 8.1 x 10E22 miles = 1.3 x 10E29 millimeters.

For comparison, the distance from Earth to the Moon is 1.2 light-seconds, and the distance to the Sun is 500 light-seconds. Planet Pluto, the most distant solar planet, is at an average distance of 39.5 AU (Astronomical Units) = 0.0006 light-years from the Sun. On the average, it would take 5.5 hours to reach Pluto by radio from Earth. Because the 500 light-second-radius of Earth's orbit is inside Pluto's orbit, the distance from Earth to any point of Pluto's orbit varies by approximately plus/minus 500 light-seconds every year.

By the "Big Bang" theory, the age of the universe is 13.7 billion years. Yes, it is the same number: 13.7 billion years of time and 13.7 billion light-years of distance!

Google: < hubble measures the expanding universe >

Note that one billion in American usage is one thousand millions = 1,000,000,000 = 10E9.

Note that one billion in European usage is one million millions = 1,000,000,000,000 = 10E12.

This book uses the American meaning for the word one billion = 10E9. It is pronounced ten to power nine, or it is number one followed by nine zeroes. (For large numbers, it is less confusing to use exponents of number ten than counting all the zeroes because the exponent-number already gives the number of zeroes.) The exponent (number nine in this case) gives the number of zeroes for one billion.

The Milky Way Galaxy is one of the 'billions and billions' of galaxies in the known universe. The total number of galaxies in the known universe is estimated to be of the order of 80 to 120 billion, or more.

Google: < how many stars in the milky way galaxy? >
          < how many stars in the universe? >
          < how many galaxies in the universe? >
          < nearest 50 stars >
          < milky way galaxy >

'Our' Sun is one star of the 100 (or 200 to 400) billion stars in this slowly rotating Milky Way Galaxy of stars, boulders, dust and gases. The Milky Way is a disk-shaped spiral galaxy with a few arms resembling a toy pinwheel. It is thicker near its center than near the outer rim. The diameter of the Milky Way Galaxy is approximately 100,000 light-years. The Sun is approximately 20,000 to 30,000 light-years from the center of the Milky Way Galaxy. The 'outside border' of the Milky Way is not exactly defined because at its outskirts it is just more or less a dense 'swarm' of stars/dust and other material.

'Our' Sun (with all its nine (eight 2006) planets and more) orbits around the center of the Milky Way Galaxy at a speed of approximately 235 kilometers (145 miles) per second going once around in 200 to 300 million years, which is called a **galactic year**. There are faster stars nearer to the center of this galaxy orbiting its central Black Hole (one or two black holes?) at speeds of at least 5,000 kilometers (3,000 miles) per second, which is more than 1% of the speed of light.

Unaided normal human eyes can usually see only those stars that are within approximately 2,000 light-years from us. However, there are some variable stars at greater distances that might be seen when they are at their brightest. Recall that our distance from the Sun is only 0.000016 light-years. The 2,000 light-year-radius sphere all around us is practically our Celestial Sphere. The nearest of these naked-eye stars is 4.24 light-years away from us. From one spot on Earth, approximately 2,000 to 2,500 stars may be visible to good naked eyes and about 5,000 to 8,000 from the whole Earth. The upper limits are for perfect human eyes on moonless clear nights. More stars can be seen from the Southern Hemisphere than from the Northern Hemisphere because the southern end of Earth's spin-axis points toward the central plane of the Milky Way disk. Telescopes can 'see' billions of galaxies and stars. Because there are 'only' 31.5 million seconds in one year, one realizes that only small astronomical samples can be studied.

To repeat:

1. This Solar System is at the center of the 4,000 light-year diameter naked-eye-Celestial-Sphere. This sphere is approximately 28,000 light-years from the center of the Milky Way disk.

2. This Solar System is approximately 60 to 70 light-years on the north side (above?) of the central galactic plane, which 'here' is approximately 200 light-years thick. The south end of Earth's spin axis points toward the galactic central plane. This means that from southern latitudes, more stars are visible than from northern latitudes.

3. As has been mentioned, this Milky Way Galaxy is only one galaxy among the 80-125 billion (or more) galaxies from here to the distance of 13.7 billion light-years. The nearest large disk-shaped spiral galaxy to us is the Andromeda Galaxy at approximately 2,300,000 to 3,000,000 light-years away from us. The diameter of the Andromeda

Galaxy disk itself is approximately 220,000 light-years long. Because this Milky Way Galaxy is rotating, it is estimated that the distance between us ('our' Sun) and Adromeda is decreasing by approximately 100 kilometers (60 miles) per second i.e. as seen from this solar system in the Milky Way pinwheel arm, which is 'spinning away' in a direction away from Andromeda, although Andromeda itself may be moving at 300 kilometers (200 miles) per second toward the center of the Milky Way. If the distance at the present time *between these two galaxies* is 3,000,000 light-years, then after one thousand years this distance will be 'only' 2,999,999 light-years. If this approaching continues, after a few billion years, Andromeda and Milky Way probably will collide like many galaxies do. As that separation decreases, the velocity toward their collision may increase because the gravitational pull between these two galaxies will be increasing. In the 'collision' of these galaxies, most stars will not collide.

In the one-millimeter 'Real Overview Model' (a few pages back), this Milky Way Galaxy and the Andromeda Galaxy are three one millimeter cubic boxes apart.

Google: **< star constellations >**
                 < m31 the andromeda galaxy >
                 < celestial coordinates >

SET THE REFERENCE POINTS ON 'YOUR ODOMETER' AS FOLLOWS:
(Forget about your car's odometer.)

Radius of Earth is 6371 km = 3959 miles.

Circumference of Earth is 40,000 km = 24,850 miles.

7.5 times around Earth's equator is one light-second.

Distance to Moon is 1.2 light-seconds.

Sun is 500 light-seconds away = One Astronomical Unit = 1 AU = one Astronomical Unit = 149, 598, 000 km = 92,956,000 miles.

Sun's RADIUS is 1.85 times our lunar distance, or 2.32 light-seconds. Sun's DIAMETER is twice that long.

Pluto (the most distant solar 'planet') is at 39.5 AU from the Sun = 5.909,121,000 km = 19,700 light-seconds = 5 hours 28 minutes (plus-minus 500 light-seconds, or plus-minus 1 AU).

Nearest star to us is at approximately 4.2 light-years away.

Almost all naked eye stars are within about 2,000 light-year distance from us.

The diameter of the Milky Way Galaxy-disk is about 100,000 light-years

We are at about 28,000 light-year distance from Milky Way's center.

Nearest spiral galaxy resembling this Milky Way Galaxy is the Andromeda Galaxy at approximately 3 million light-years away from us

Most distant known (2007) galaxies are at 13,700,000,000 (13.7 billion) light-year distance from us.

These distances can describe all distances used in Earth and Space Sciences.

13. Earth's Orbit around 'Our' Sun

'O ur' Sun and the orbits of all of its nine planets can be considered to be just a point at the center of the 4,000 light-year diameter naked-eye-Celestial-Sphere. This Earth is at a distance of one Astronomical unit from 'our' Sun and the most distant solar planet (2006?) Pluto is 39.5 Astronomical Units from the Sun. One light year is 63,000 Astronomical Units long, and 4,000 light years is approximately 253,000,000 Astronomical Units long.

Google: **< the nine planets >**
              < one light year distance >

Earth is orbiting 'our' Sun at an average distance of one Astronomical Unit = 149,600,000 kilometers = 93,000,000 miles = 500 light-seconds in the plane of the ecliptic (Ecliptica) with an average linear speed of 20 miles (30 kilometers) per second.

Earth's orbital plane is called the **plane of the Ecliptic**, or **Ecliptica**. The shape of the orbit is slightly elliptical with a very small eccentricity of 0.0167. A perfect circle has an eccentricity of zero.

Our minimum distance of 0.98 Astronomical Units to 'our' Sun happens when Earth is at its Perihelion point (nearest point of the orbit ellipse to the Sun) in the early part of January. Our maximum distance of 1.02 Astronomical Units to 'our' Sun happens when Earth is at its Aphelion point (farthest point of the orbit ellipse from the Sun) in the early part of July.

Earth's orbital speed is slightly faster in January than in July, but still very close to the mentioned 20 miles (30 kilometers) per second. Earth's orbit obeys Kepler's Laws.

With respect to 'our' Sun, Earth is repeating its same annual orbital ellipse year after year with very small and very well-known variations. These variations will not be covered in this book.

Google: < kepler's laws >
           < planetary orbital elements >
           < jpl solar system dynamics
           < ellipse – from mathworld >
           < precession and nutation >

If the stars could be seen during the day, an observer on Earth would see 'our' Sun's position change/move in Earth's orbital plane against the ('fixed') star background by approximately one degree per day or about two Sun's diameters per day. 'Our' Sun could be compared to the center of a merry-go-round, which seems to cover all compass-directions on the merry-go-rider's horizon. This apparent movement of 'our' Sun by approximately one degree per day against the star background is important when understanding how time is defined and measured.

Google: < precession and nutation of the equinoxes >

As Earth orbits the Sun, it keeps its N-S spin axis tilted to its orbital plane in a fairly constant manner by 23.5 degrees. The small movements/gyrations of the spin axis are discussed later in connection with the Precession and Nutation.

The gravitational attraction of 'our' Sun keeps this Earth in its orbit. Without the Sun's gravitational pull, Earth would fly tangentially fairly straight into very cold oblivion of the surrounding space in this Milky Way Galaxy. The intensity of the Sun's gravitational attraction (pull) here at Earth's distance at one AU distance (1 AU = 149,600,000 kilometers = 92,960,000 miles) from 'our' Sun is approximately 596 milligals, or about 0.0006 G of Earth's average gravity of 981,000 milligals = 1 G. The Sun's attraction keeps all nine planets, including Earth, solar asteroids and comets in their orbits around 'our' Sun. All objects in this solar system are captives of 'our' Sun's gravitational pull.

Google: < fundamental physical constants > When there, click on 'Adopted
                                        values', and then 'standard ac-
                                        celeration of gravity'.

Actually it is the center of mass (some call it center of gravity) of the Earth-Moon-system, the **Barycenter**, which orbits 'our' Sun along the pretty Keplerian orbit ellipse and not the center of Earth. The instantaneous Barycenter is inside Earth on the straight line connecting the centers of the Earth and Moon. This point inside Earth moves around approximately 1000 kilometers (600 miles) from the center of Earth and is always on the straight line connecting Earth's center to the Moon's center. Therefore, the center of Earth wobbles in and out of the Keplerian orbit. **Repeating: it is actually the Earth-Moon Barycenter that traces Earth's Keplerian solar orbit.**

At the time of full moon, the center of Earth is 'inside' the orbit ellipse. At the time of a new moon, Earth's center is 'outside' the orbit ellipse.

Kepler's Second Law of planetary motion states that the mentioned radius vector joining Earth to the Sun sweeps over equal areas in equal times as Earth travels along its orbit ellipse at any time of the year. For instance, in one day the mentioned radius vector sweeps a triangular sector with its central angle a little greater in January than in July, but **the areas per one day (or any given constant duration of time) of the orbital sectors are constant.** The sectorial angles for equal time durations are a little larger in January than in July, but the radius vectors are correspondingly shorter in January than in July to keep the areas constant.

Google: **< kepler's laws of planetary motion >**

So, the radius vector from the center of 'our' Sun to Earth (1 AU long straight line from the Sun to Earth) sweeps and defines our orbital plane once around (360 degrees in 365.242 days) in one year, i.e. it sweeps almost one degree arc (0.986 degrees) of Earth's orbit each day. As was mentioned, this orbital plane, the ecliptic plane (Ecliptica), is a very stable plane year after year with respect to 'our' Sun.

'Our' Moon can be up to approximately five degrees (5.1454 degrees = inclination of Moon's orbit to the plane of Earth's orbit) 'above', or 'under' the ecliptic plane. Because the ecliptic plane itself is tilted by approximately 23.5 degrees from Earth's Equator, the Barycenter (inside the Earth) can be at latitudes between 28.5 degrees North or/and South of the Equator. (There are also other barycenters. For instance, the Sun-Earth system has its own barycenter.)

Then one has to consider also that during the time from new to full moon, Earth spins (rotates) almost 14 times around its axis, when 'our' Moon makes only one half of its orbit around Earth. As mentioned, the Moon's orbital plane is 'tilted' (inclination) by approximately five degrees to the ecliptic plane. Therefore, the distance of the Barycenter from Earth's center also varies by a small amount because the Moon's orbit is slightly elliptical with an eccentricity of 0.0549.

All of this is too complicated for this book. The following Internet sites will somewhat clarify the situation.

Google: **< the ellipse >**
                 < ellipse calculator >
                 < orbit of moon around the sun >
                 < barycenter of moon and earth >
                 < earth's seasons >

Eons ago the Creator made and put together a rather complicated orbital situation. Men from Copernicus, Galileo Galilei, Kepler, Newton, Einstein and many others have applied science to these heavenly 'bodies'. Today's scientists with their modern telescopes, instruments and computers are able to observe, measure and compute the mentioned heavenly movements and details with great accuracy.

We should remember that the Creator made it all to be 'just so', and that we are free to make our observations, measurements, computations and deductions keeping in mind that **there is very little all mankind can do to change or even tweak anything in God's marvelous designs.**

Earth's orbit ellipse rotates (think about the major axis of the ellipse) once around 'our' Sun in the ecliptic plane in approximately **26,000 years,** which amounts to about 50 arc seconds per one year. **This motion is called precession,** which is similar to the gyration of a fast spinning toy top on a smooth tabletop. This 26,000-year gyration is caused mainly by the gravitational attractions of the Moon and Sun on Earth's ring-like equatorial bulge, which is not a point mass, nor a spherical mass. The north end of Earth's spin axis points nowadays within one degree of arc to Polaris, 'our' North Star. In 26,000 years, Earth's spin axis gyrates once around in the sky making a full circle with approximately 23.5-degree radius around the pole of the ecliptic plane. The pole of the ecliptic plane is 90 degrees from the plane of ecliptic. Therefore, in 26,000 years Polaris will be once again the North Star. After 13,000 years Polaris will be 47 (two times 23.5) degrees away from the North Celestial Pole. By definition, the North Celestial Pole is the point where the instantaneous spin axis points.

Google: **< precession of earth >** Scroll down and read about the obliquity of the ecliptic plane.

The 'vacuum of space' around 'our' Sun, where Earth orbits, is not a perfect vacuum. **As this Earth orbits the Sun, every day/hour/minute/second there will be small incoming meteorites (with their own velocities) hitting Earth's atmosphere raining meteor dust down to Earth by** approximately **40,000 – 150,0000 – 300,000 metric tons per year, or about 100 – 400 – 800 tons per day.** The incoming rocky/metal meteorites are heated by the air's ram pressure to glowing temperatures. They will become visible as the 'shooting stars'. The ice in the incoming particles melts and vaporizes. The incoming masses are insignificant amounts when compared to Earth's total mass, which is 5.972 x E21 metric tons. Very seldom do bigger meteorite chunks hit the Earth. The Sun is also spewing out small particles for the Earth to plow through (solar wind) at speeds of 300 – 800 kilometers = 200 – 500 miles per second. Space weather will not be described in this book.

Google: **< today's space weather >**
               < milky way >
               < meteor dust >
               < 10-accretion of mass > Scroll down to see an astronaut on the Moon.
               < meteorite odds/ends & trivia >
               < solar wind >
               < current solar wind conditions > Try to get this type of up-to-date daily information from any printed book without current Internet sites!

14. CREATOR, GOD, SUPREME BEING

There seems to be a general tendency to avoid words 'Creator' and 'God' in some discussions about the universe, solar systems, planets and the existence of life. **Those words cannot be completely avoided and should not be avoided.** They are intertwined with the universe and the human mind. **A common question seems to be: if God didn't do it all, who or what did? Nobody here on Earth knows for sure! Let each one of us have our own beliefs. All of those opinions cannot be correct anyway!** None of us, and neither did our ancestors do much of anything in the area of creating anything from nothing! There is endlessly much more to learn about the universe, 'our' Solar System, this Earth and life. Earth is just a little spec in the Solar System, although it is the whole world for our lives and existence.

Most of us are deeply grateful for our opportunity to live and to have a large number of choices, opportunities and possibilities in our lives on this **unique** and wonderful **planet Earth.**

It is scientifically known fact that there is no other place (planet, moon, asteroid) in 'our' Solar System where human life is possible without life-support systems taken there from Earth. This Earth is also the only place in this solar system where the birds sing, where green grass grows and where one can smell roses.

The nearest star with possible Earth-type life is more than half a dozen light years away. **Physical travel there would take many centuries and four times as many human generations. It would take a similar duration of time for the Aliens to travel here.**

Therefore, all stories about Aliens on Earth in the past, present or future are just unverified imaginary stories.

15. Number of Galaxies and Stars in the Universe

No one knows very well the actual number of various types of galaxies in the known universe with its radius of 13.7 x 10E9 light-years (2007), and with its volume of approximately 1.08 x 10E31 cubic light-years. Many estimates exist for those numbers, although the number 13.7 billion seems to have stabilized in recent years. The space around us continues to the limitless infinity.

Google: < how to calculate the volume of a sphere? >
                 < power of ten multiplier chart >

Estimates for the total number of galaxies in the known universe seem to range from 80 billion to 120 billion and even up to 3,500 billion galaxies. **This range prorates to one galaxy per 1.35 x 10E20 to 3.1 x 10E18 cubic light-years**.

One way to estimate the number of galaxies in the universe is to use its total volume and extrapolate from the number of known galaxies in nearby cubic light-years with the understanding that the density of galaxies in the universe is varies..

For example, the Andromeda Galaxy is within **three million light-year** distance from us. The volume of this sphere (containing Andromeda) is approximately 1.13 x 10E20 cubic light-years. Many Internet sites list about 30 galaxies within this 3 million light-year radius from us. **This prorates to** approximately **one galaxy per 3.9 x 10E18 cubic light-years = one galaxy per 3.9 cubic-million-light-years.**

If these numbers will be extrapolated for the whole universe, there would be 2.8 x 10E12, or **2,800 billion galaxies in the whole known universe.** This 'rhymes' rather well with the **estimated numbers of 80 to 3,500 billion galaxies in the whole universe.**

Google: < how many galaxies are there in the universe? >
             < list of nearest galaxies >

The 2,800 billion estimate obtained by extrapolating from our nearby galaxies seems to be a reasonably good 'ballpark estimate'. **Note that all these numbers are estimates.**

An average galaxy contains from 100 to 400 billion stars. Multiplying the number of galaxies (2,800 billion) by the average number (300 billion) of stars in each, **the total number of stars in the known universe by this estimation is 8.4 x 10E23, or of the order of 10E24.** There are other similar 'ballpark' estimates for the total number of stars in the known universe. The exponent 24 may be somewhere between 20 and 25.

Google: < how many stars in the universe? >

'Our' Milky Way Galaxy is just one galaxy among all galaxies in the known universe. The Milky Way Galaxy is estimated to have from 100 to 400 billion (10E11 to 4 x 10E11) stars.

One of the Milky Way stars is 'our' Sun. Earth is one of the nine (eight in 2006) **major planets orbiting 'our' Sun. Of these nine solar planets, this Earth is the only one where humans can exist** without taking along their own breathing air and temperature controls.

Voila, here we are, finally!

The numbers above are enormously large numbers. For instance, **think of the enormity of the number of 10E13 (= estimated number of galaxies in the known universe) in the following way.** Since one year has about 31,556,928 seconds, then 317,000 years are needed to have 10E13 seconds.

One could cover all those 10E13 galaxies by devoting one second of studying time per each galaxy (each galaxy with hundreds of billions of stars) studying day and night continuously for 317,000 years. We simply cannot know everything! One cannot learn very much in one second. Astronomers are giving us good samples and estimates.

We are truly fortunate that we are Earthlings and have numerous possibilities on how to live our life spans here on Earth. **The Creator/God has been**

exceedingly generous and good to us. Those who resent or don't believe in the existence of God or a Supreme Being ought to be grateful to whatever they believe is responsible for the past existence of their ancestors and finally responsible for the existence of themselves.

16. There is Nothing in the Known Universe Like this Earth

'**O**ur' **Sun has nine planets. Of these nine, only this Earth is suitable for human type life**. For any type of life, this Earth is definitely 'it'.

The **inner planet Venus** is much too hot for humans with its crushing atmospheric pressure. In addition, it does not have liquid water or breathable air. Being there would be much worse than being in a pressure cooker here on Earth.

Google: < **planet venus statistics** > Scroll down to see in the table.

The table shows that the rotational period -243.0 days and the orbital period +224.7 days are almost the same. One can wonder if the absolute values of these two numbers are slowly becoming the same due to the Sun's tidal pull on Venus resembling 'our' Moon, whose spin and orbital rates around the Earth are the same. The minus sign in front of 243.0 indicates that Venus is rotating in a clockwise manner. This Earth rotates counterclockwise as seen from space above the Earth's North Pole.

The other **inner planet Mercury** is much too hot on its dayside and much too cold on its night side for humans with no atmosphere for breathing. Temperatures of the ground over there swing 1.100 F = 600 C degrees in one Mercury's day (= 58.6 Earth days) between +810 F and -360 F degrees = between +470 C and – 180 C degrees. Water on its surface would soon boil into very hot steam, which would then escape into Mercury's surrounding space.

Google: < **periodic table: lead** > Lead melts at 327.5 °C = 600.65 °K = 621.5 °F.
    < **planet mercury statistics** > Scroll down to the table.

The **outer planets Mars, Jupiter, Saturn, Uranus, Neptune and Pluto** are much too cold for human type life. Water on their surfaces would freeze to ice in a short time.

Google: **< outer planets >**
                < the nine planets >

This Earth is in the 'sweet', Goldilocks-type spot in this solar system; it is not too close, nor too far from 'our' Sun. In many respects, Earth's atmosphere and one G-gravity are also ideal for us all.

Even 'our' Sun is in a 'sweet', Goldilocks-type spot in this Milky Way Galaxy; it is not too close to, nor too far from the galactic center.

'Our' Sun's distance from the Milky Way galactic center at 28,000 light-years is ideal. Much closer than that to the Milky Way center, the cosmic radiation would be too strong and the greater star density and star interactions would be too strong for comfort and lasting life. The outer edges of the Milky Way generally don't have many solar systems like this one, mainly because heavier molecules 'out there' are scarce.

This Earth is by far the best and the only known planet (or moon) **in this Solar System capable of supporting human life**. The surface of Earth is truly a wonderful place for us to live. Here on Earth there are thousands upon thousands of systems/things and their combinations, which combine to make our lives possible. **This Earth is absolutely unique in the known universe.**

This Solar System is so well known that it can be said with certainty that **nothing else similar to this Earth exists in this Solar System.**

Further, nothing similar has been found in any other 'nearby' solar system so far, and the same goes for the rest of the universe.

Google: **< views of the solar system >**

However, this does not mean that some other Earth-like, life-supporting planets cannot exist somewhere outside this Solar System. This disk-shaped-spiral Milky Way, our home galaxy has 100 to 400 billion (100,000,000,000 plus) stars. One of them is 'our' Sun.

The Milky Way Galaxy **has 'only' millions of stars (not billions), which could have Earth-like planets** orbiting them. All such eligible stars are from 8 to 70,000 light-years away; therefore, they are 'way out' of reach for us Earthlings, except for possible radio/laser contacts. Numerically one-way radio/laser transmission travel times in years, are the same as the distances in light-years.

A laser/radio message takes 8 years to travel an 8 light-year distance to the **nearest possible planetary system**, and the same time is needed for a possible reply. How about two-way communications over 100 light-year distances? Another hurdle is to be able to send strong enough signals to reach the intended destination.

There are a few stars nearer to us than 8 light-years, but it is believed that they cannot have Earth-like planets mainly because they are too cold or double stars (doublets), and also for some other known reasons (radiation is one them). A life-supporting star must have friendly radiation, and the life supporting orbiting planet cannot be too close nor too far from it.

Google: < **star alpha centauri** >
< **nearby stars** >
< **the 50 nearest stars** > Scroll down to the table of stars.
< **milky way galaxy** >

The possible **existence of Earth-like planets, or any life supporting planets in other galaxies, which are at distances of millions of light years away from Earth, will remain eternally unknown to us Earthlings**. Due to the great distances, there is absolutely no hope of finding any Earth-like planets in any other galaxy. **Even a two-way radio communication would take over five million years with anything in Andromeda Galaxy, There are a few smaller galaxies nearer than Andromeda, but even they are hopelessly too far for any communication with Earthlings.**

Until 2007, over two hundred planets orbiting some of our 'nearby' Milky Way stars have been found. They have been Jupiter-size/type large gas-giant planets too far from their suns to support human type life. **However, there might well be Earth-like planets nearer to some of those suns, so there is some hope for radio/laser contacts with the possible dwellers of those planets in the future. A two-way radio/laser communication would take more than fifteen years.** Talk about delays between back-and-forth conversations!

Those other solar systems so far found, where Earth-like planets may exist, are at distances not much greater than five thousand light-years from us. Only a relatively 'small' number of all Milky Way stars (suns) can be studied for extra-solar planets. Stars on the other side of the Milky Way disk at some 50,000 light-year distances are blocked out by the Milky Way's center.

Google: < seti >

17. This Solar System in the Milky Way Galaxy

Our Solar system (Sun, nine – eight since 2006 – planets, moons around planets, asteroids and comets, which orbit the Sun) **is in one of the spiral arms of the** 100,000 light-year diameter **Milky Way Galaxy.** Distance from here to the center of the galaxy is approximately 25,000 to 30,000 light-years.

It is believed that the galactic center contains a super massive black hole (or two) with a mass of over 3 million times the mass of 'our' Sun. Some stars near the galactic center are orbiting it at speeds in excess of 5,000 kilometers (3,100 miles) per second (1.7 % of the speed of light). Some of them have only 15-year orbital periods around the Milky Way center. Compare that to our galactic year of over 200 million years!

'Our' Sun (with its planets and moons) takes 225 million years (one galactic year) to orbit once around the galactic center at a speed of approximately 235 kilometers (150 miles) per second. The central portions of the Milky Way 'spin' faster around the galactic center than the outer parts

Google: < milky way galaxy >
             < galactic year >

17.1. Distance Unit Parsec

On some of the Internet sites and elsewhere, a distance unit of one parsec is used instead of one light-year. One parsec is 3.086 x 10E13 kilometers = 1.918 x 10E13 miles = 3.26 light-years.

One parsec is defined to be that distance from where one Astronomical Unit (1 AU = 149,600,000 kilometers = 92,956,944 miles) can be seen in an angle of one arc-second.

Prorating this to a more comprehensible scale, recall that one-millimeter can be seen in an angle of one arc-second at a distance of 206 meters = 676 feet.

One parsec distance is not 'that much' different from a light-year; therefore, it is not used as much as the light-year.

18. ANGULAR UNITS OF MEASUREMENTS

18.1. RADIANS

In a circle, select a piece of its circular arc, which is as long as the radius of the circle. Connect each end of this arc to the center of the circle with radii. The angle between these two radii is one radian = 57.3 (57.29577951) degrees of arc = 206,265 (206,264.8062) arc-seconds.

Recall that if a chord distance in a circle between two points on its arc is as long as the radius of the circle, the angle from the end points of the chord to the center is 60 degrees.

Google: **< radian measure >**

18.2. DEGREES, MINUTES AND SECONDS

The total circumference of a circle (once around) is 360 degrees = 24 hours = 2 x Pi radians = 6.28 radians = 1,296,000 arc-seconds. One arc-degree is = 60 arc-minutes = 3600 arc-seconds. Many theodolites can measure angles with one arc-second precision.

18.3. HOURS, MINUTES AND SECONDS

As everyone knows, the Earth rotates around its North – South spin axis 360 degrees in 24 hours, which prorates to 15 degrees per one hour resulting in 24 time zones (24 x 15 = 360) on the Earth. In some applications (not in this book) hour angles are in use. So, once again: 360 arc degrees = 24 hours of time.

In some applications, as stated above, it is useful to know that one arc-second is an angle in which one millimeter can be seen from a distance of 206 meters (676 feet). Of course, it takes a good telescope and a stable atmosphere to be able to read millimeters at a distance of 677 feet (206 meters).

Google: **< coordinate systems in astronomy >**
             < surveying instrument collection – theodolites >
             < kern geodetic theodolites >
             < Surveying, Engineering, & Construction Instruments >

19. Sun and its Planets

The mass of 'our' Sun is 1.989 x 10E27 metric tons, or 330,000 times the mass of this Earth. Earth's mass is 5.976 x 10E21 metric tons, including all of us and all our belongings. Sun's equatorial radius is 695,500 km = 432,200 miles = 2.32 light-seconds and its diameter is 4.64 light-seconds = 4.64 ls.

Some Milky Way stars have smaller and some have larger masses than 'our' Sun. Some stars are brighter; some are dimmer than 'our' Sun. Of all the Milky Way stars, only 10 %, i.e. over 8 to10 billion stars, have a greater masses than our Sun.

Google: < galactic black holes >
          < the nine planets > Click on the Sun
          < the sun and its solar system >
          < centrifugal and centripetal forces >
          < sun's rotation period >
          < the very latest soho images >

http://sse.jpl.nasa.gov/planets/profile.cfm?Object=Sun&Display=Facts&Sy
    stem=Metric
http://www.space.com/php/multimedia/imagegallery/igviewer.
    php?imgid=4162&gid=298

The diameter of 'our' Sun is 1,391,000 kilometers (864,300 miles) = 4.64 light-seconds or approximately 190 times as long as the Earth's diameter. **The Sun's radius is 1.8 as long as the distance from here to 'our' Moon.** Imagine that! The average distance from here to 'our' Moon is 60 Earth's radii = 1.2 ls.

The surface temperature of 'our' Sun is approximately 5,800 Kelvin = 5,500 Celsius = 10,000 Fahrenheit degrees. Sun's interior is the millions of degrees **hot**.

The Sun is a very hot ball of gas. Its rotation period varies with latitude since it is made of gas. It rotates in a counterclockwise manner when viewed from celestial North Pole. Its equatorial regions rotate their 360 degrees faster than its polar regions. The equatorial regions rotate once around in approximately 26 Earth days. The regions at 60 degrees of Sun's latitude rotate once around in about 31 Earth days. Polar regions rotate once around in approximately 36 Earth days.

20. Comparing the Gravitations of the Sun, Earth and Moon

We live and move under the influence of 1 G gravity conditions 24/7 every second of our lives. Some airplane rides, amusement park rides and vehicles can subject (omit violent collisions) their occupants for short durations to gravity values from zero G to 2 G or 3 G, seldom higher. Some airplane pilots are tested for short durations to approximately 9 G. The human body cannot tolerate 9 G, or greater Gs for extended periods. Only Earth orbiting astronauts and a few Moon visitors have experienced near zero Gs for a few days/months.

The intensity of gravity (gravitation) on the Sun's surface is approximately **27.5 G,** or 27.5 times stronger than here on Earth. Earth's average gravity is 1 G = 981 centimeters per second-squared = 981 gals = 981,000 milligals = 32.2 feet per second-squared.

1. **One-meter (39 inch) drop of a pen here on Earth, where the prevailing gravity is one Gal = 1 G:**
   If a pen is dropped from one-meter (39-inch) high table to the floor (in air, not in water) here on Earth under the influence of 1 G gravity, the **pen hits the floor 0.45 seconds later at a speed of 10 MPH** (16 kilometers per hour). If a person falls from that table to the floor, she/he also will hit the floor at 10 MPH unless the hit is slowed down by hands/arms/legs.

2. **A one-meter (39 inch) drop under prevailing 27.5 G acceleration is achievable in a centrifuge simulating the Sun's gravity:**
   If the same **pen is dropped under the influence of 27.5 G acceleration** from a one-meter (39 inch) 'high' support to the 'floor', which

in a spinning centrifuge will be the outside wall, the pen would be 'sucked' to the centrifuge wall in 0.09 seconds. **It would hit the centrifuge wall at a speed of 51 MPH** (82 kilometers per hour).

3. **A one-meter drop under the influence of the Moon's gravity of 162.3 gals = 0.165 G:**
   If the same **pen is dropped** from one-meter (39-inch) high support to the 'ground' on the Moon, the pen would hit the 'ground' 1.1 seconds later at a speed of 4 MPH = 6.5 kilometers per hour.

Human beings could not live under 27 G acceleration conditions for many seconds. The 27 G gravitation (= acceleration) would make the person weigh 27 times her/his normal weight, and it would flatten any person against the ground (centrifuge wall) more than ten times harder than any wrestler could. Talk about being breathless!

Breathing and heart beats would stop pretty quickly under 27 G acceleration conditions. Much of the muscle-mass and body-tissues would be pulled from the top of the body around the sides of the body 'down' to be nearer to the 'ground' (centrifuge wall), and the skin from the top of the body would be stretched over the sides of the body. **That is what 27 G gravity would do to your dead body!**

One G gravity here on Earth (outside spinning centrifuges) is just about right for us.

When an airplane takes off from an aircraft carrier deck, the pilot experiences about 3.3 G acceleration to reach a speed of 180 miles (290 kilometers) per hour in 2.5 seconds on a 100-meter (330-feet) long runway. It is 'quite a jerk' for those 2.5 seconds that no one wants to experience over extended periods.

21. Incoming Meteor Particles and Dust to Earth

The material objects (masses) in this solar system are: the Sun, the nine (eight in 2006) planets, their numerous moons and millions of smaller objects called asteroids/comets, and of course, since October 1957, the minute man-made artificial satellites orbiting the Earth. The Sun contains more than 99.8% of the total mass of the Solar System, leaving less than 0.2 % for all other listed objects.

Together with the nine solar planets, most of the asteroids and comets do not get very far 'above or under' the Earth's orbital plane, i.e. their orbital inclination angles from Earth's orbital plane are only a few degrees. Their orbit ellipses can reach over 50 times as far from the Sun (50 AU) as the Earth (1 AU) does.

Google: **< inclination for solar orbits >** Then click on Solar System Data to see the inclination (Incl) column.

When in outer space, large orbiting objects are called **asteroids/comets.** All masses in this solar system orbit something else; there are no stationary objects out there! When comets are not very far from the Sun, **they have a dust tail and an ionized tail** generally pointing away from the Sun. **Asteroids do not have an appreciable tail.** When small asteroid, comet or debris particles enter Earth's atmosphere, they are called **meteoroids.** Their sizes vary from one millimeter to several kilometers (miles). If they land on Earth (on water or land), they are called **meteorites.** A **meteor** is the flash of light (shooting star), not the particle itself. Small incoming particles (cosmic dust) may vaporize/burn/break to pieces in Earth's atmosphere depending on their composition, i.e., ice, fragile stone, more durable iron or other material.

Google: < the nine planets solar system tour >
                < web definitions for radiation pressure >
                < solar sails, latest news >
                < asteroids, comets, meteoroids, meteorites, meteors >
                < artificial satellites >
                < solar wind >
                < comet tails >

http://antwrp.gsfc.nasa.gov/apod/ap061004.html
http://antwrp.gsfc.nasa.gov/apod/ap060319.html
http://cfa-www.harvard.edu/iau/lists/InnerPlot.html

Asteroids and comets are orbiting 'our' Sun according to Kepler's laws. Millions of them are between the orbits of the planets Mars and Jupiter and millions more orbit beyond Pluto's orbit. From time to time, some of those asteroids/comets can come close to other asteroids/comets. 'It' can get fairly crowded out there, which can be seen from the pot marks on many photographed asteroids. In densely populated asteroid/comet belts, orbits can get perturbed (deflected, distorted) by constantly changing gravitational attractions (interactions) between many 'nearby' asteroids. The perturbed orbits may become more flattened (stretched out) ellipses so that their perihelion points (nearest orbital points to the Sun) will then be inside Earth's orbit. After all that, this Earth may have a brand new visitor.

Some comets have hyperbolic orbits. After 'swinging' once around the Sun, they will never return to this solar system for another visit. One recent comet with hyperbolic orbit was comet McNaught at its perihelion in January 2007. Another recent comet was SWAN with orbital elements:

Calculations by B. G. Marsden on July 21, July 28 and August 4, Marsden published an orbit on September 21, which revealed that the comet moved in a hyperbolic orbit, as follows:

"The following orbital elements are taken from *MPC* 57794:

C/2006 M4 (SWAN)

Epoch 2006 Sept. 22.0 TT = JDT 2454000.5

T 2006 Sept. 28.7283 TT MPC

q 0.783022 (2000.0) P Q

z -0.000302 Peri. 62.5930 -0.2221106 +0.8475967

+/-0.000032 Node 148.7267 +0.1501991 -0.4586221

e 1.000236 Incl. 111.8227 +0.9633832 +0.2669185

From 106 observations 2006 July 12-Oct. 5, mean residual 0".6.

The listed eccentricity e of SWAN's orbit e = 1.000236 indicates that the orbit is a hyperbola. (If the eccentricity e = 0, the orbit is a perfect circle; if e < 1, the orbit is an ellipse; if e = 1, the orbit is a parabola; if e > 1, the orbit is a hyperbola.)

The listed inclination i = 111.8227 indicates that SWAN's orbital plane is very steep with respect to the ecliptic plane (= Earth's orbital plane and also close to the orbital planes of other solar planets). Therefore, comet SWAN's orbit will not be appreciably perturbed by other solar planets, and it will continue on its way into the Milky Way Galaxy, never to return to this solar system again, as earlier mentioned.

Comet McNaught (January 2007) was so very bright because its perihelion passage was only at 0.17 AU = 25,432,000 km = 40,928,000 miles. Comparing this to planet Mercury's average distance from the Sun = 0.38 AU, one finds by the 'Inverse Square Law' that the solar radiation upon comet McNaught at its perihelion passage was approximately five times stronger than on the sunny side of Mercury, which is hot enough to melt tin.

Google: < one astronomical unit >

    < asteroids and comets >

    < comet plunging into the Sun >

http://www.gsfc.nasa.gov/topstory/20011025comet.html

http://marine.rutgers.edu/mrs/education/class/paul/orbits.html

    < orbital mechanics >

    < Solar radiation – Encyclopedia of Earth >

http://www.spacescience.com/newhome/headlines/ast03nov99_1.htm

http://science.nasa.gov/headlines/y2006/13jun_lunarsporadic.htm

http://www.spaceweather.com/comets/gallery_mcnaught.htm

http://www.spaceweather.com/comets/gallery_cometswan.html

Another source for a few near-Earth asteroids is the Moon's ground surface. Some pieces of the Moon's ground surface have been splashed (sent flying) from many high speed (from 10 to 70 km/s = from 6 to 45 miles/s) asteroid/comet crash sites at velocities greater than the Moon's escape velocity of 2.5 km/s = 1.5 miles/s. After such an impact, there will be new pieces/boulders orbiting the Moon/Earth system. From the Earth those incoming pieces/boulders look like regular asteroids. However, this situation is the so-called unstable 'three body problem' with its three bodies being: the ejected boulder, the Moon and Earth. Quite likely, sooner or later the boulder will have a crash with the Earth or the Moon. This has happened many times. Due the vacuum around the Moon, **the ejected dust flies just once settling down** – if it does not fly away faster than the Moon's escape velocity.

The same applies also to the planet Mars when asteroids/comets hit it (escape velocity of Mars is approximately 5 km/s = 3 miles/s). Thousands of Moon surface asteroids have been found in/on the Antarctic ice. **It is obvious that when 70 km/s = 45 miles/s asteroids hit the Moon or planet Mars, some pieces will fly off at speeds greater than the corresponding escape velocity.** Dust storms have camouflaged many impact craters on Mars but not on the Moon.

Google: < lunar impact craters >
            < the three body problem >
            < searching antarctic ice for meteorites >

If a standard military rifle is fired on the Moon, the bullet will land somewhere on the Moon's ground with almost its initial muzzle velocity of one kilometer (0.6 miles) per second. If a bullet is fired on the Moon with a muzzle velocity greater than Moon's escape velocity (2.5 km/s = 1.5 miles/s), that bullet may enter into its own orbit and may even land here on Earth as a meteorite.

Google: < principal lunar craters >
            < antarctic meteorites >
            < escape velocity of the nine planets >

The 2003 Chicago 'rock' meteorite was stony and approximately 6 feet in diameter. It broke into several pieces over the Chicago area.

Google:　　< chicago 2003 may 6 meteorite >
　　　　　　< large asteroids >

About ten objects of this size enter Earth's atmosphere every year. Tons of smaller pieces hit the atmosphere every day.

Google:　　< meteor showers > Click on 'Calendar' and 'Periodic'.

According to some estimates 100 to 100,000 metric tons (estimates vary that much!) of asteroid material falls to Earth every day. This amount prorates to 6 milligrams to 6 grams per person per one year assuming the world population to be 6 billion. One cubic millimeter raindrop with a diameter of 1.24 millimeters weighs one milligram. Once in awhile, we are breathing a few molecules of meteor dust among other dust particles!

Prorating the 100 tons for smaller areas, one gets the following approximate amounts. (For 100,000 tons, multiply by one thousand.)

One square mile receives approximately 500 milligrams per day. Compare this to six 81-milligram Aspirin tablets.

One acre receives 0.8 milligrams (one square mile = 640 acres) per day.

One square kilometer receives approximately 200 milligrams per day.

One hectare receives approximately 2 milligrams per day.

The total amounts of dust, comet and asteroid pieces and the comets and asteroids themselves existing in the solar system are continuously getting reduced as these masses cease to exist as separate masses by their collisions with the Sun, planets and moons. For this reason, this solar system is being swept cleaner all the time. There may be changes coming to this cleaning process, as explained below.

Particles of asteroids and comets hit the Earth's atmosphere (orbiting with the Earth) almost continuously on a daily basis. (The orbital velocity of the Earth is 30 km/s = 19 miles/s). The debris does not just sit there waiting for the Earth to 'come and hit them'. The clouds of debris are in their own elliptical solar orbits following roughly the nucleus (head) of their parent comet/asteroid at speeds **up to 42 km/s = 26 miles/s.** Some debris crosses the Earth's orbit every day of the year. If the Earth happens to go through that part of its orbit, there will shooting stars.

After hitting the atmosphere, the small debris pieces float down as dust. The 'medium-size' particles will hit the ocean/ground usually at velocities of less than one Mach (= velocity of sound at sea level). Bigger pieces come in faster.

An asteroid 50 meters (150 feet) across collides with Earth once in approximately 600 years. Seventy percent of them will hit an ocean and thirty percent will hit a land area. Large asteroids may hit the ground almost vertically like the Barringer asteroid did in Arizona about 50,000 years ago, or they may come in slantingly grazing the Earth/atmosphere like the Tunguska asteroid did in 1908 over Siberia.

Large asteroids like the Barringer asteroid will not be slowed down very much by the atmosphere before hitting the Earth's surface unless they break into many smaller pieces. The impact velocities with the ground can be from 10 to 50 km/s = 6 to 30 miles/s because large asteroids will not be slowed down by very much during their last couple of seconds of their final flight.

Google: **< mach number >**
                < tunguska asteroid >
                < barringer asteroid >

This solar system, situated in one of the pinwheel arms of the Milky Way Galaxy, is orbiting the center of the Milky Way with and within the arm according to Kepler's Laws in complicated ways because there are many masses all around in the 'neighboring' stars and dust clouds with some boulders in them. All those masses have their own gravitational pulls in their own directions.

The 'nearby' Milky Way dust/debris/stars in this same Milky Way arm are also orbiting at their own speeds either a little faster or slower than what we are. As a result of these speed differences, it is believed that right now we are entering a somewhat denser galactic cloud volume – maybe for some thousands of years to come – than where this solar system has been recently. There is no reason to be concerned about our safety due to this natural development.

The surroundings of many young Milky Way stars are 'dirtier' with more dust and boulders than what this over four billion-year-old solar system planetary disk has. Solar systems get somewhat 'cleaned out' of their dust and debris during the millions/billions of years as the collisions with the larger objects terminate the existence of the smaller 'stuff'. This solar system is much cleaner of debris/comets/asteroids than what it was eons ago.

Google: < asteroids and comets >
            < shoemaker-levy asteroid home page >
            < images of the moon >
            < planet mercury >
            < the planetary society headline for 06/10/98 >

21.1. Incoming Asteroids and Debris to Earth's Vicinity

As was mentioned, in the distant past there was more dust, more asteroids and more comets in this solar system than at the present time. Their existing numbers are continuously being reduced by their crashes into the Sun, planets and moons. Many photographs of the moons, Mercury and Mars show that the solar planets/moons have been bombarded by thousands of large asteroids in the past. Many of the very old craters don't show well (or at all) on the Earth's oceans (71 % of Earth's surface), on the rainy and windy continents, on the cloud-covered Venus, on the windswept Mars, or on the big gas planets and, of course, on the Sun itself.

Google: < the nine planets >

The big craters on 'our' Moon, Mercury, Venus and Mars are mostly from collisions eons ago. Many large and small asteroids and comets have been and are continuously being destroyed by diving into 'our' Sun, planets or into their moons.

Fortunately large asteroids hit the Earth very seldom anymore. The intervals between large hits have been on the order of many thousands/millions of years.

In addition to the mentioned three sources, **asteroids can come dangerously close to this Earth also from a fourth source.** All possible sources are listed below.

1. From their regular elliptical solar orbits.

2. **From their disturbed/deflected solar orbits by the gravitational pulls of their nearby occasional orbital objects.** Asteroids in their normal (not dangerous to Earth) solar orbits can be deflected to new and dangerous orbits by the gravitational pulls of other asteroids, and/or the gravitational pulls of the planets Mars, Jupiter and other planets in many progressive/successive and accumulating ways.

3. Asteroids may be pieces ejected from the Moon or Mars from large asteroid collisions.

4. **Asteroids can come from outside this solar system** from the unknown parts of the interstellar space in the Milky Way Galaxy or possibly from the intergalactic space outside the Milky Way Galaxy, which can be considered 'far fetched' but possible.

The 'regular' near-Earth elliptical solar orbits of the asteroids/comets are not eternally constant. Their newer deflected orbits may lead them to crash into Earth, Venus, Mercury or the Sun. More likely they will continue to fly by those two planets and then past the Sun. At their new perihelions (nearest point to the Sun) they have their maximum orbital speeds, and then they start slowing down as they start 'climbing up higher' from the Sun in their altered orbits. On their way toward their new aphelions (farthest point from the Sun) they will have brand new opportunities to fly by or crash into the solar planets/moons. When they are 'going away' from the Sun at one Astronomical Unit distance, their speed is again up to approximately 42 kilometers (26 miles) per second. After many years they are coming back for their next 'laps' around 'our' Sun. Then they have again new chances for encounters with the Sun and solar planets/moons.

During the past eons, thousands of comets and asteroids in their orbits around the Sun have crashed with solar planets and their moons, and many more have crashed into the Sun, never coming close to the Earth again. There have been pictures taken and published on the Internet of comets just about ready to dive into the Sun and Jupiter.

Google: < comets hitting the sun >
               < comet shoemaker – levy 9 >
               < asteroids and comets >
               < welcome to the planets >
               < the nine planets >
               < the sun and its solar system >
               < centrifugal and centripetal forces >
               < milky way galaxy >

The Internet site 'the nine planets' has a wealth of information about all solar planets, most moons, asteroids and the whole Solar System.

21.2. COMET TEMPEL-TUTTLE'S DEBRIS AND LEONID METEOR SHOWERS

Comet Tempel-Tuttle was discovered and so named in 1865. Its orbit ellipse is over 20 Astronomical Units long (= major axis of its orbit is 10 AU long). Its whole orbit long tail has **caused annually (at least until 2004) the Leonid meteor showers in the latter part of November**, when Earth zips through its orbit ellipse. The Tempel-Tuttle swarm of meteor particles is mostly small ice/dust/gravel/rock debris in the comet's whole orbit-long elliptical tail. The comets nucleus sweeps inside Earth's solar orbit, when it is near its perihelion (nearest to the Sun). The latest perihelion passage of the comet's nucleus was on February 28, 1998. At its aphelion in 2014, the comet's nucleus will be outside Saturn's orbit at approximately ten Astronomical Units from the Sun, where its orbital speed is the slowest. One could say that the orbital speed is the slowest when the object is at its 'maximum altitude (height) above the Sun', from where its speed starts to increase as it falls to 'lower altitudes above the Sun'. When crossing Earth's orbit, Tempel-Tuttle's speed is approximately 40 kilometers (25 miles) per second.

Goggle: < comet tempel-tuttle, the leonid comet >

In November 2003, five years after the latest 1998 perihelion nucleus passage, the nucleus was beyond the orbit of Mars on is way to its aphelion, and

Earth simultaneously (in November 2003) went through a part of the comet's wispy tail. This gives an idea of the length of the comet's tail.

Cometary Tail: http://www.ifa.hawaii.edu/~jewitt/tail.html
http://www.spaceweather.com/comets/gallery_mcnaught_page12.php
To enlarge, click on thumbnail photos.

The duration of the observed Leonid meteor showers around November 19, 2003 was approximately one day. In that one day Earth traveled about 2.5 million kilometers, or 1.6 million miles, or approximately 7 times the Earth-to-Moon distance. This is a good sample of how thick Tempel-Tuttle's tail was in that region, which Earth traversed in November 2003.

The velocity of Earth is 30 kilometers (19 miles) per second, and the Leonid-particles travel in almost opposite direction at approximately 40 kilometers (25 miles) per second, so the mutual collision speeds are about 70 kilometers (43 miles) per second. Most particles float down as dust, burning up in the atmosphere. Bowling ball-size pieces hit the ocean/ground at speeds of 80 to 100 meters (250 to 300 feet) per second, or up to approximately 200 MPH. **The result of such a hit to the trunk of a car can be seen on the following Internet site**.

Google: **< peekskill fell october 9, 1992 >** Scroll down on 'Peekskill photo' to see a rocky piece of the meteorite that hit the car trunk.

When the meteor pieces hit the Moon (they do), the speeds given above are the impact speeds pulverizing/melting whatever they hit. If the pieces hit planet Mars (they do), they are not slowed down very much by the thin Martian atmosphere. The incoming particles may glow, but not burn, from the 'air' friction before hitting the Martian 'ground'. There have been some reported flashes on Mars. They could be flashes of light on the surface when a sizable meteoroid hits the ground, and the kinetic energy converts to heat energy.

As was mentioned, the Tempel-Tuttle comet together with its whole-orbit-long tail is in a 'very' elliptical solar orbit. Tempel-Tuttle's year, or its one complete 360-degree orbit around the Sun, is 33 Earth-years long. The debris particles are strewn behind the nucleus of the comet, and they essentially cover/trace the whole huge orbit ellipse behind and now even ahead of the comet's nucleus. The orbital wispy debris cloud is like an elliptical wreath in the sky. **Because the cloud is not uniform, the number of 'shooting stars' from the Leonids varies from year to year.**

The tail of a comet can be compared to a non-uniform, spotty and very wispy and thin contrail of a high-flying jet. Every year Earth gets one sample of the debris-cloud in the month of November as Earth zips through the comets fast moving tail. The little particles of the comet are few and far between. Due to the high speeds of the particles and Earth, when (for instance) two pea-size particles cause two consecutive visible bright streaks in the sky one second apart, quite likely those two particles were 70 kilometers (45 miles) apart from each other. **This debris cloud seems to be in a fairly stable orbit, because it is in Earth's 'way' every November like clockwork.** Because the Tempel-Tuttle's period is 33 years, the same bunch of debris will be in Earth's 'way' once again in approximately another 33 years. **In the other years, Earth gets other new samples of the wispy long debris cloud.** The number of debris particles in a comet's orbit is reduced somewhat every year when the Earth and Moon wipes a few thousands of them away.

One of the 'Leonids samples' will be of special interest, when **the nucleus itself comes close to Earth among its own debris field next time in 2031.** Even then, there is lots of room in the space around Earth's orbit for a safe passage by the nucleus instead of having a major collision with Earth. Astronomers will publish information about the comet's approach in due time.

There are several other annual meteor showers somewhat similar to the Leonids.

Google: **< annual meteor shower information >**

When comets with large orbital eccentricities (very flat elliptical orbits around the Sun) travel near Earth, here at one Astronomical Unit (149,600,000 kilometers, 93,000,000 miles) from the Sun, their speed is approximately 42 kilometers (26 miles) per second, which is obtained by multiplying Earth's orbital speed 30 kilometers (18.6 miles) per second by the square root of number two according to Kepler's Laws.

Where does the 'square root of two' come from? According to Kepler's Laws, the 'square root of two' is actually for parabolic orbits, but for very flat elliptical orbits it is a good approximation. Therefore, **according to Kepler's Laws there are no asteroids/comets in any *solar* orbit, which in the vicinity of Earth (at one AU distance from the Sun) has greater speed than the mentioned maximum speed of 72 kilometers (45 miles) per second with respect to the Earth.**

When there is a fly-by of **any object traveling faster than 42 kilometers (26 miles) per second here at one AU distance from Sun, that object will never come back for another visit because it cannot be in an elliptical solar orbit like good comets and asteroids are.** Such fast movers can come from outer space, where speeds of 300 kilometers (200 miles) or more per second are not uncommon. When/if these fast moving objects are in our neighborhood, they will just fly by this whole solar system only once. They will never return for another visit. Their speed exceeds the escape velocity of this solar system. The Milky Way Galaxy has had many star explosions, and pieces from such events can fly in many directions for thousands/millions of years at high velocities.

Google: **< escape velocities in celestial mechanics >**

Earth's atmosphere brakes the asteroid/comet dust particle speeds down to snowflake-speeds or to a couple of hundred meters (feet) per second speeds, depending on the mass/density/shape of the incoming object. Very massive incoming objects may not slow down very much before hitting the ground. Large asteroid/comet hits to Earth are very rare; maybe there has been one large one in approximately one thousand years. Bowling ball-size objects hit Earth more frequently. About 70 % of the incoming asteroids/comets hit the oceans, and most of those go unnoticed, since only about 30 % of Earth's surface is land area.

Earth's artificial satellites are in their orbits outside most of the atmospheric air at distances from a few hundred kilometers (miles) to about 42,200 kilometers (26,200 miles) and beyond from Earth's center. Many of them have been 'nicked', and solar panels have been punctured by asteroid/comet particles traveling at relative speeds up to 72 kilometers (45 miles) per second. At satellite altitudes there is practically no atmospheric braking to slow down the incoming particles. Dents in the satellite bodies and holes in the solar panels by small asteroid/comet pieces have been confirmed by astronauts visiting the Hubble Telescope.

Google: < meteoroid hits to hubble telescope >

21.3. ASTEROIDS FROM INTERSTELLAR AND INTERGALACTIC SPACE

Interstellar space is the space around/between the stars in the Milky Way Galaxy. **Intergalactic space** is the space between galaxies.

Many Milky Way stars have exploded, and the boulders, dust, gas and other debris are flying away from the explosions for eons at great speeds in many directions. The hyperbolic orbits (with respect of this solar system) of the 'incoming' asteroids can be in any plane when approaching this solar system. They may be deflected into new 'flying' directions by the gravitational pulls of 'our' Sun, a planet, or a moon. They may fly through this solar system disk at steep angles (inclinations) at great speeds only once never to return for another visit. Wherever they go after their one-time fly-bys is not very important.

There can be one-kilometer (mile) size or larger boulders (asteroids) (any size will do) **coming this way from parts of the Milky Way Galaxy from outside this solar system. Most of them will just sail harmlessly by us or go through the entire solar system disk between the orbits of planets.** Some of them will pass or go through the solar system disk without human awareness.

At a speed of 'only' 300 kilometers (200 miles) per second, one light-year distance is traveled in one-thousand years. At that speed the traveling times from the nearest stars are of the order of 4,000 to 5,000 years.

Google: < extrasolar asteroids >
             < asteroid belts >
             < jpl solar system dynamics >
             < kinetic energy >

It is certain that **there have been many more asteroid fly-bys than colli-sions**. Most of the fly-by asteroids have missed 'our' Sun, Earth and other solar planets and their moons.

There have also been some collisions, and a few big asteroids have collided with Earth. For instance, there is a crater in Mexico **112 miles (180 kilometers) wide and believed to be 3,000 feet (900 meters) deep. The crater was probably created by a large asteroid collision** approximately **65 million years ago.** It could have come from outer space or from the asteroid belt. Who knows? Large asteroids have also hit Earth at many other places and other planets/moons thousands of times.

The Barringer meteor with a 150-foot (50-meter) diameter hit Earth in Arizona approximately 50,000 to 65,000 years ago. Its estimated incoming speed was 18 kilometers (11 miles) per second. At those speeds a large asteroid goes through the Earth's atmosphere in a few seconds.

There are many visible meteor craters on 'our' Moon, Mercury and on many planetary moons.

Google: **< revolution and rotation of the planets >**
                < asteroids from outer space >
                < history of asteroids from outer space >
                < moon's geological history >
                < planet mercury >
                < tunguska asteroid >

When a high-speed incoming *asteroid from outer space* is approaching 'our' Sun, it will most often be deflected by the Sun's gravitational pull (attraction) **into a never-to-return hyperbolic orbit** as far as the Sun is concerned. Of course, it could also dive into the Sun or much less likely hit a planet or a moon.

When an asteroid *from this solar system* is approaching its Sun, it will most likely not hit any planets or the Sun. It will just continue for another solar orbit.

For an asteroid to be able to collide with a planet, the asteroid has to travel almost in the plane of the Ecliptic (=solar disk). For a possible collision, its perihelion point (nearest point to 'our' Sun) must be inside the orbit of that planet. Even then, the planet may be far away from the point where the asteroid crosses near the planet's orbit.

Asteroids coming from outside of this solar system will *most likely* be traveling in some plane other than that of the Ecliptica. In that case, they will most likely go through the whole solar system disk without a solar/planetary collision.

An **incoming asteroid with a perihelion distance less than one astronomical unit nearly in the plane of the Ecliptic may be difficult to detect in the following special situation.** If such an asteroid is approaching 'our' Sun, and if it barely misses the Sun, it may be coming toward the Earth almost **straight from the direction of the Sun if Earth happens to be at 'that' time (of the year) at 'that' point in its orbit where the asteroid crosses Earth's orbit. There may be a close call or a collision with little warning.**

Note that there were many 'ifs' in that particular situation for a possible collision. Fly-bys are much more common than collisions.

It may take an extra-solar fast moving asteroid only 5-6 days from its perihelion point to arrive in the neighborhood of Earth's orbit at one AU (astronomical unit) **distance from the Sun**. Typical velocities of the interstellar boulders are approximately 300 kilometers (200 miles) per second, which is about ten times faster than Earth's speed in its orbit.

If the asteroid is orbiting near Earth's orbital plane, and if its perihelion point is within 1 AU (Astronomical Unit) from **the Sun, it may cross Earth's orbit at two places or near to them. Earth is a moving target in its orbit for incoming asteroids.** Earth may be far away from those two points of danger. **In just two months Earth travels one Astronomical Unit distance in its orbit. There can be a collision only if Earth happens to be close enough to one of those two danger spots in its orbit,** *where and when* **the asteroid will cross Earth's orbit.**

All other solar planets will also most likely avoid collisions with the incoming asteroids from outside this solar system. Those incoming asteroids most likely will go through the solar disk of planets from either 'upside' or from 'downside' if it can be agreed which side of the solar disk (Ecliptica) is up or down.

Where is 'up' and where is 'down' from the solar disk (ecliptic plane) of planets? If the Australians will agree, it could be said that 'up' is on the side where the North Star Polaris is.

For example, consider Earth (or any other planet or moon) in its regular orbit is 'getting in the way' of an incoming asteroid/comet/etc. as they approach this solar system from far away places. When they approach a certain spot in Earth's orbit-ellipse at one particular time, Earth may be far away at some other part of its orbit, and nothing noteworthy happens. But **sometimes (very seldom) the asteroid/comet can come so close to Earth that Earth's gravitational pull 'sucks' it in for a collision.**

Consider that an **incoming** comet/etc. comes fairly close to Earth (or close to any other planet/moon) but not close enough for a collision. **By coming close to Earth, Earth's gravitation will deflect the comet's orbit into a new direction; thereafter, it will have a brand new orbit. New encounters may/will follow later** with several bodies in the solar system.

Recall that this Earth is moving at 30 kilometers (20 miles) per second in its annual orbit around the Sun. When those *solar system* comets/etc. in their highly elliptical solar orbits are at 1 AU distance from the Sun (1 AU = Earth's distance from Sun), they move at speeds of *up to* 42 kilometers (26 miles) per second. Therefore, the maximum collision speed with Earth can be approximately 72 kilometers (44 miles) per second, and the minimum collision speed is approximately 12 kilometers (7 miles) per second. **When an extra solar system asteroid approaches, its speed can be much higher.**

21.4. Annual Meteor Showers

This solar system is *mainly* a Pluto-orbit size circular disk with a radius of approximately 40 astronomical units. (One AU is the distance from here to the Sun = 500 light-seconds = 149,600,000 kilometers = 92,956,900 US Statute miles). As was mentioned, most of its mass is in the Sun and its nine planets, asteroids, comets and dust/gravel/debris. All the nine solar planets and the millions of asteroids, numerous comets and the dust and debris have their own more or less circular or elliptical orbits, *mostly* located very closely in the same plane where the Earth orbits. Earth's orbital plane is 'our old friend', the very steady plane of the Ecliptic (Ecliptica).

Leonid and some other *annual meteor showers* appear regularly in Earth's atmosphere at certain months and days of a year. The Earth goes through 'the Leonid-sector' of its orbit every November. All the gravel/dust/debris is in its own elliptical solar orbit (whole orbit long elliptical comet's tail), which intersects Earth's orbit at the area where Earth is at the latter part of November. The speed of the Leonid debris pieces themselves, when crossing Earth's orbit, is approximately 40 kilometers (25 miles) per second. Recall that the Earth itself is zipping along at 30 kilometers (20 miles) per second.

Recall that the debris is not just sitting at any one place in space. If there would be a stationary debris-mass/cloud or any other mass anywhere in this solar system, it would start at once to 'drop' toward the Sun or a nearby planet, and a new Keplerian orbit would be born for each stationary debris particle.

Saturn's moon Enceladus is orbiting Saturn in its own comet-tail-like debris field tail called Saturn's E ring. Many comets have orbit-long debris-tails, also.

Google: < annual meteor showers >
http://antwrp.gsfc.nasa.gov/apod/ap070327.html This is a photo of Enceladus
in Saturn's E ring.

The elliptical orbit of the *annual meteor shower debris* (annual for Earth; the debris is orbiting constantly for eons in their own annual orbits) is composed of more or less dense, wispy cloud-puffs of debris, widely strewn over the total length of the comets elliptical orbit up to several Astronomical Units long. Like clockwork, every year when Earth goes through the same part of its orbit (November for the Leonids), there are some new streaming debris particles for Earth to 'plow' through.

Every year there is a new portion of the debris stream crossing Earth's orbit, which was not 'there' one year earlier. Also, every year for the duration of the comets orbital period, Earth gets a 'new dose' of the particle stream. For Leonid meteor showers, this happens at the end of November (November 18-19 in 2003) when the Earth in its orbit is approximately 45 degrees 'shy' from its January perihelion passage.

21.5. COMETS AND LARGE ASTEROIDS

According to most reasonable estimates **1,000 to 10,000 metric tons of meteorite material from surrounding space floats down to the oceans and the ground every day.** On the average this is from **0.7 to 7 milligrams per one year on each square meter (yard)** of Earth's surface. Recall that one cubic millimeter of water weighs one milligram.

Only very small amounts of normal house dust may be fresh from the outer space. The dust, which collects on shelves, tabletops and on other surfaces, is mostly just dust carried around by winds from some other locations on Earth. Particles from some dust storms in the Sahara are blown across the Atlantic Ocean to America.

According to some estimates, **there are half a million asteroids** (many are pieces of rock/ice) **with masses of more than 300 million tons each** in solar orbits between distances from 2.2 to 3.3 AU from 'our' Sun. There are some other asteroid belts at about 50 Astronomical Units from the Sun. The vast majority of them are safely (for us) orbiting the Sun between the planets Mars and Jupiter not far from the plane of the Ecliptica (= plane of Earth's orbit). There are also some with eccentric orbits that can and do come inside Earth's orbit. Of course, that could become dangerous for life on Earth if the asteroid and Earth happen to be near the same part of Earth's orbit at the same time. **Only a very few asteroids can be threats to Earth.** Many of those asteroids have craters/indentations/nicks on themselves from collisions with other asteroids.

Google: < The Landscape on Comet Tempel 1 >
              < asteroids and comets >
              < near earth asteroid tracking >
              < near earth object program >
              < the threat and promise of asteroids and comets >
              < astronomy picture of the day archive >
              < nasa image of the day gallery >. Also click on 'View Archives'
                                    under the small image. Click
                                    on image to enlarge it.
http://www.universetoday.com/am/publish/bang_asteroid_hit.html

Among the larger objects is the recently discovered Quaoar, which has a 620-kilometer (390 mile) radius. It is at a distance of 42 AU from 'our' Sun. Recently, there have been some discussions of possibly including some of the more or less irregular orbiting bodies into the count of solar planets, making the total count to twelve from the traditional nine. Whatever the objects are called, they are a part of this Solar System for eons to come.

Google: < quaoar >

The space between the orbits of Earth and Mars is estimated to contain over one thousand orbiting one kilometer (0.6 miles) size asteroids, and maybe over a million 50 meter (150 feet) size rocks, all traveling at their normal orbital speeds. Fortunately, collisions by large asteroid pieces are extremely rare.

Google: < 1908 asteroid > This site describes the 1908 Siberian asteroid.

Pictures of 'our' Moon, planet Mercury and even some larger asteroids show numerous craters/indentations created by collisions with smaller objects. Many asteroids and comets also dive into the Sun, Jupiter and Saturn without leaving lasting visible indentations. Recall from earlier that some of the orbits of incoming asteroids/comets can be perturbed (altered) by gravitational attractions of nearby planets/moons to new orbits. Those incoming pieces which hit the Sun, planets or moons certainly will not hit this Earth after that. **This is another neat feature God/Creator made for our safety and wellbeing.**

Small objects in space, including 'our' Moon, don't have an atmosphere because their gravitational attraction is too weak to keep gas molecules around them even if they had some atmosphere to begin with.

The great *majority* of asteroids never come close to Earth. Some asteroids have eccentric orbits, which can take them to within one Astronomical Unit of the Sun (where we are). Thus, they may cross Earth's orbit coming uncomfortably close to us when Earth and that asteroid happen to be near the same spot in Earth's orbit at about the same time. In such a case Earth's gravitational pull will deflect the asteroid into a new orbit, or the asteroid may be sucked into a collision.

Usually Earth is 'somewhere else' in its orbit, when an asteroid/comet 'comes around'. If it passes close to, say, Saturn, Jupiter or Mars, it may be deflected by any of them into a brand new orbit, which has never before been observed. It may also crash into one of those 'arrow catcher' planets/moons like thousands of them have done before, or the object may crash into Earth

Similarly, an asteroid/comet approaching Earth may be deflected by the Moon's gravitational pull and then by Earth's gravitational pull into a new solar orbit. If so, then the comet/asteroid will have brand new possibilities for encounters/crashes with Venus, Mercury and 'our' Sun itself on its way to perihelion (nearest orbital point to the Sun). After going around the Sun, it will have opportunities for crashes/new deflections by Mercury, Venus and Earth and the outer planets again into another new orbit. Collisions and near misses are extremely rare but they do happen.

Large incoming asteroids will not be slowed down much by Earth's atmosphere.

'Our' Moon, Mars, Venus and Mercury show much evidence of being good 'arrow catchers' for us. Other solar planets and the Sun itself have the same 'arrow catching' function for our safety and benefit. It is obvious that after an asteroid/comet collides with something much bigger than what it is, it ceases to be dangerous to anything else.

After a long-lasting asteroid/comet is deflected many times and enough, it may crash (and many do crash) into the Sun. **The SOHO satellite has photographed such final events for several comets.**

Google: < the moon fact sheet >

                 < asteroids and comets >

                 < the solar and heliospheric observatory >

                 < the very latest soho images > SOHO = The Solar and Heliospheric Observatory.

                 < asteroid collisions with earth >

                 < meteoroids and meteorites >

                 < photos of asteroids > To enlarge an image, click on it. Note that even small asteroids have been 'nicked' by other asteroids.

21.6. ASTEROIDS/COMETS/DEBRIS APPROACHING 'OUR' MOON

'Our' Moon also orbits through most of the same widely spread wispy debris clouds causing the Leonid and other meteor showers on Earth. **The debris clouds/gravel/rocks orbiting close to 'our' Moon can also hit Earth's atmosphere 3 to 4 hours later if the Moon happens to be on the side of the incoming particles to Earth. If those particles hit Earth's atmosphere, they may become 'shooting stars' of various sizes.** If the Moon is on the other side of Earth in its orbit, the comet/meteor debris will encounter the Moon some 3 to 4 hours later after the particles went by us.

Those particles that barely miss 'our' Moon are deflected into brand new orbits ('up', or 'down', 'left', or 'right') as the Moon goes through the meteor debris stream. In this fashion the meteor stream gets 'unfocused/scattered around' during the years, and the cleaning/spreading out process of meteor showers from the solar system continues.

Those particles, which hit the Moon, hit its surface with their orbital velocities increased slightly by the Moon's gravitational pull (attraction). Even small grains of debris/sand/whatever hitting at those extremely high speeds can have devastating effects for anything on the Moon. During a period of a month those incoming particles can hit the Moon's surface from almost any horizontal compass-direction and from many altitude angles.

The speed of the debris particles near 'our' Moon can be up to 40 kilometers (25 miles) per second. Recall that for the incoming particles from outside the Moon's orbit, the Moon in its monthly orbit is only 1.2 light-seconds ahead or behind the orbiting Earth, or its solar distance is only 1.2 light-seconds greater, i.e. 501.2 light-seconds at full moon, or 1.2 light-seconds lesser, i.e. 498.8 light-seconds at new moon than Earth's distance from the Sun (= one Astronomical Unit = 500 light-seconds).

The Moon's orbital velocity around Earth is 1 kilometer (0.62 miles) per second, and together with the Earth, its average annual velocity is the same as Earth's annual orbital velocity = 30 kilometers (18.6 miles) per second. At the time of a full moon, the Moon's velocity with respect to the Sun is approximately 31 kilometers (19.3 miles) per second and at the time of a new moon approximately 29 kilometers (18 miles) per second.

The following **example is about the Leonids and the Moon**. Around the latter part of November, if the Moon happens to be on the incoming side of the Leonid-stream particles, some of these particles either hit the Moon or fly-by toward Earth. The incoming speed of the Leonids is 40 kilometers (25 miles) per second almost head-on with Earth. These speeds combine to encountering speeds of the Leonids with the Earth and Moon to approximately 70 kilometers (43 miles) per second.

There are no 'shooting stars' on the Moon, since there is no braking/slowing-down atmosphere around it. Instead of braking, the Moon's gravity pulls and sucks the incoming particles in for collisions with even faster accelerated speeds. The Moon's ground surface has continuously (more so in the eons past but still continuing today) been peppered to fine powder (regolith) by those high-speed, high-kinetic energy particles hitting the Moon's ground at velocities from about 10 kilometers (6 miles) per second to 40 kilometers (25 miles) per second. Those velocities are 10 to 40 times greater than the muzzle velocity of a bullet from a military rifle. When the velocity increases by a factor of 10 to 40, kinetic energies for a given moving mass increase by a factor of 100 to 1,600.

During the eons past, some larger asteroids hitting 'our' Moon have splashed some pieces of its ground so fast and far away that Moon's gravitation was not strong enough to pull all the pieces back to it. The Moon's escape velocity is about 2.4 kilometers (1.5 miles) per second. Any piece flying faster than that away from the Moon will not fall back. Some of those escaped pieces have been found in Antarctica.

Google: < regolith on moon >
< Boeing Celebrates the 30th Anniversary of Apollo 11 >
  Scroll down to see a footprint.
< neil amstrong's footstep on the moon >
< lunar meteorites >
< definition of escape velocity >
  http://antwrp.gsfc.nasa.gov/apod/ap051217.html
http://science.nasa.gov/headlines/y2005/07dec_moonstorms.htm?list186835

Over the eons some of the ground on 'our' Moon has been pulverized to regolith by millions and millions of incoming meteorites/asteroids evident from the moon-astronaut's footprints.

Judging from the photographs taken of the footprints of the Moon Astronauts, it seems that the ground is loose, fine, powdery material (regolith) worked over and over from the rocky Moon's surface during the past eons by incoming high-speed meteorite particles. The almost constant pounding, rock breaking/sand-blasting process is continuing even today. As the incoming meteorites hit, their high kinetic energies are converted to heat. Many resulting light flashes have been photographed from the Earth. The sudden heating partially melts some regolith together with the incoming meteorites at the target spot to glass like pieces. At some places on the Moon the regolith is ten meters (30 feet) deep. The regolith dust just sits there because there is no wind to blow it around.

Google: < **lunar regolith breccias** > Scroll down to see the footprints and the 'peppered' pot-marks around them. Clicking on some pictures will enlarge them.

To see the devastation on the far side of the Moon, go to the Internet site: **http://antwrp.gsfc.nasa.gov/apod/ap070225.html**

As the photograph shows, **Moon's far side has been crashed into and peppered by asteroids/meteorites much more severely than its near side.** One explanation for the difference in the number of Moon's craters is the following:

1. Consider the Moon's monthly orbit around the Earth with its orbital plane within plus-minus 5 degrees 'above-and-under' of the plane of Earth's orbital plane = the ecliptic plane = Ecliptica = (Zodiac-plane).

2. Most of the asteroids/meteorites coming to the vicinity of the Earth and the Moon are also orbiting in their own planes not far 'above-or-under' the ecliptic plane.

3. Consider that **one swarm of meteorites/asteroids** is coming toward the Earth at the time of a full Moon **from behind the Moon**.

4. Consider that **simultaneously a similar swarm of meteorites/asteroids** is coming from behind the Earth (as seen from the Moon) heading toward the full Moon.

5. In the case of 3 the **Moon is acting as 'an arrow catcher' for the Earth** and many asteroids/meteorites in the swarm will get their first chance to crash into the far side of the Moon leaving fewer meteorites/asteroids to come to the Earth's neighborhood some 3 to 5 hours later.

6. In the case of 4 the **Earth is acting as 'an arrow catcher' for the Moon** and all asteroids/meteorites in the swarm will get their first chance to crash into the Earth's atmosphere **leaving many fewer pieces of the swarm** to approach the near side of the full Moon some 3 to 5 hours later.

7. Under these circumstances, the far side of the Moon gets more hits than its near side. The photographs show the results on the following Internet sites:

   http://antwrp.gsfc.nasa.gov/apod/ap070225.html
   http://www.nineplanets.org/luna.html
   http://ase.tufts.edu/cosmos/view_picture.asp?id=524

When looking at the photo of the astronaut's footprint of the large shoe/boot of the astronaut, and considering that:

a) the astronaut with his gear might weigh 300 pounds (136 kilograms) on the Earth, and that

b) gravity on the Moon is approximately one sixth of Earth's gravity, then the situation is about the same as when a 50-pound (22-kilogram) boy on Earth wearing those big shoes/boots would step into similar powdery (talcum like) material.

What kind of dust must the loose powdery material (regolith) be for the boy's 50-pound weight on Earth to create the pictured imprint? It would have to be the consistency of dry wood-ashes, dry wheat flour, talcum powder, or something similar. This will give an idea of the consistency of the regolith material on the Moon.

Some indentations seen on the Moon's ground next to the astronaut's footprints were made by one millimeter, some by one centimeter and some by an inch size particles and bigger **hitting the ground at speeds about 30 times faster than the proverbial rifle bullet.** If one of those holes in the ground would appear next to an astronaut, he would not hear the zip or the thump, but he could feel the ground disturbance with his feet.

Millions of micrometeorites strike the Moon's surface and also Earth's atmosphere every day with similar density of particles. Moon's surface is as dangerous place for visiting astronauts as an open rifle firing range here on Earth. The entire Earth's surface would be equally dangerous if it were not for our protective atmosphere, where most of the incoming pieces burn to dust floating slowly and safely down to the oceans and the ground. This is another obvious proof of the wisdom and protective care of our Creator for all life here on Earth, and it is another good reason to be grateful for our safety. Big asteroid hits, as seen in the Moon photographs, especially on its far side, happen seldom but they do happen.

21.7. ASTEROIDS AND COMETS HITTING 'OUR' MOON

Most of the asteroid/comet pieces are small, only from dust particles to pea-size; only seldom are they bowling ball size or larger. **Due to the lack of atmosphere around 'our' Moon, hits are at full speed whether the incoming piece is a dust particle or one kilometer (mile) or more in size. When an asteroid/comet hits the Moon, it does not slow down at all, but it is only accelerated to even faster speeds by Moon's gravitational pull.**

The fact that the Moon does not have an atmosphere is due to its weak gravitational attraction. The total mass of 'our' Moon is not large enough to keep possible atmospheric gas molecules from escaping to the emptiness of the surrounding vacuum of space. Therefore, **when an asteroid/comet hits the Moon, the maximum impact speed can be faster than 72 kilometers (45 miles) per second.**

Why so fast? It is because the Moon orbits with Earth 30 kilometers (19 miles) per second around the Sun, the Moon's orbital speed around Earth is one kilometer (0.6 miles) per second and the solar system asteroids/comets can travel

at up to 42 kilometers (26 miles) per second at one AU (Astronomical Unit) from
the Sun. Adding the speeds together, one gets the maximum possible speed to be
approximately 73 kilometers (45 miles) per second, plus the extra accelerations by
the Moon's gravitational pull.

 • Even small particles at these high speeds can be deadly to astronauts on the
Moon. Several flashes of small meteorites hitting the Moon have been photographed after the Leonids in November 2006.

In these considerations, the **average distance of the Moon from Earth's**
center 384,400 kilometers (238,900 miles) **is insignificant when compared to
one Astronomical Unit:** AU = 149,600,000 kilometers (92,960,000 miles) =
500 light-seconds. The light-second distance numbers show the same thing: distance from here to the Moon is 1.2 light-seconds and the distance from here to
the Sun is 500 light-seconds.

Google: **< moon >**
            < lunar impact craters >
            < ccd images of the moon > This Internet site shows many lunar
                                     craters.

http://science.nasa.gov/headlines/y2006/13jun_lunarsporadic.htm
http://science.nasa.gov/newhome/headlines/ast03nov99_1.htm
http://science.nasa.gov/headlines/y2006/28apr_skyisfalling.htm?list186835

21.8. DEADLY METEOR SHOWERS ON THE MOON

A given area on the Moon's surface gets hit in equal times by approximately
the same amount of mass consisting of asteroid/comet-debris as a similar area gets
here on Earth. So the approximate amount of the incoming mass per acre/hect-
are/square mile/square kilometer on the Moon is approximately the same as on
Earth. The Leonid meteor shower is one example.

One square mile receives approximately **500 milligrams per day.** Compare this mass to one 500 mg Vitamin C tablet, or compare it to one half of a
cubic centimeter (0.06 cubic inches) of ice, which has a mass of 450 milligrams.

One acre receives approximately **0.8 milligrams in one day** (one square mile = 640 acres).

Google: **< how many acres in a square mile? >**

One square kilometer receives approximately **200 milligrams per day.**

One hectare receives approximately **2 milligrams per day.**

The big difference between the debris/comet/asteroid particles hitting planet Earth or the Moon is that there is no atmosphere on the Moon to slow down/burn/break up the incoming pieces. On Earth most of the incoming mass floats down to Earth as dust. The incoming particles hit the Moon at their unhindered average speeds of 20 to 70 kilometers (12 to 45 miles) per second.

Recall that a mass of material moving at 30 kilometers (19 miles) per second has 900 times greater kinetic energy than the same mass of material leaving a military rifle at 1 kilometer per second.

This means that on the average, one 10-milligram (compare to 81 milligram Aspirin tablet) **incoming debris piece hitting the Moon's surface has** approximately **the same kinetic energy as a 150-grain (9.7 gram) bullet coming from a military rifle.** From the incoming amounts of mass given in the listing above, one can see that **10 milligrams of debris material hits a 12 acre-area (5 hectares) of the Moon's surface every day.** A larger 500-milligram particle has 50 times greater kinetic energy than a 10-milligram particle at the same velocity. For another comparison, a small cube of ice with its side of 2.5 millimeters (0.1 inch) has a mass of 13 milligrams. Of course, **the incoming particles come in many sizes** – pun is intended. Small incoming particles are naturally more numerous than bigger particles.

The future Moon Astronauts must take into account the incoming debris on their extended stays on 'our' Moon. **Due to the high speeds of the incoming particles there, the kinetic energies of the impacts are easily much greater than those of bullets coming from military rifles.**

On November 7, 2005, a 12-centimeter-wide (4.7 inch) meteoroid slammed into the Moon at a speed of 27 km/s = 17 miles/s. The blast had an energy equal to approximately 70 kg of TNT =154 lbs of TNT.

Google: < meteors and meteor showers >
              < armagh observatory leonid meteors >
              < jpl solar system dynamics >
              < shooting marbles at 16,000 mph >
              < nasa – fireball sightings >nasa – fireball sightings

http://science.nasa.gov/headlines/y2005/22dec_lunartaurid.htm?list186835
http://science.nasa.gov/headlines/y2006/13jun_lunarsporadic.htm
http://www.space.com/scienceastronomy/061201_moon_impacts.html
http://www.spacescience.com/newhome/headlines/ast03nov99_1.htm
http://science.nasa.gov/headlines/y2007/23jan_ltps.htm?list186835

21.9. ASTEROIDS/COMETS HITTING PLANET MARS, OTHER PLANETS AND THEIR MOONS

Planet Mars and even its two little Idaho-potato-shaped moons have been cratered by asteroids/comets. The atmosphere of Mars is so thin that the incoming bowling ball-size pieces probably hit the ground almost at full space-speed. As previously mentioned, **'our' Sun and all its nine** (eight in 2006) **planets act as 'arrow catchers' for us on Earth and also for each other.**

As has been mentioned, in addition to the nine planets there are millions of other solar orbiters. One recently discovered, relatively large object has been named Quaoar. Its diameter is 1250-kilometers (775 miles). Its average distance from the Sun is 42 Astronomical Units (42 times as far from the Sun as we are). Traveling time of light from the Sun to it takes 42 x 500 seconds = 5 hours 50 minutes. Solar radiation intensity on Earth is approximately 1760 = 42 x 42 times stronger than at Quaoar's distance.

Google: < the moons of mars >
           < planet mars >
           < planet mercury >
           < planet venus >
           < quaoar >

21.10. ASTEROIDS/COMETS/DEBRIS COMING TO THE VICINITY OF THE SUN

For good numerical data and pictures go to the following Google's Internet sites:

Google: < **asteroids and comets** > click on "Asteroid Fact Sheet", and "Comet Fact Sheet".

< **the nine planets solar system tour** > Click on "Orbit Diagrams and Distribution Graphs" and "Comets and Asteroids" and "Meteor Streams".

< **jpl** >

At those sites, one can see that even **most** listed comets have small **inclinations** of only a few degrees. There are a few exceptions, for instance the Hale-Bopp comet has an inclination of almost 90 degrees, meaning that its orbit is almost perpendicular to the solar system disk.

Inclination of the plane of the orbit = the dihedral angle of the orbital plane with the plane of the Ecliptica (Earth's orbital plane). By convention, the inclination is a number between 0 and 180 degrees.

The intersection of the equatorial plane and the orbital plane is a straight line, which is called the **line of nodes**; ascending (=rising) node is the point where the object crosses above the Ecliptica and the descending node is the point, where the object crosses below the Ecliptica.

From the list of comets and asteroids one can see that most listed asteroids have **orbital periods** of less than five years, and that the listed comets' orbital periods are from 3.30 years to 40,000 years.

One can also see that **inclinations of the nine** (eight, 2006) **planets** are small angles except for the small planet Pluto's orbit (17.1 degrees); therefore, the mass distribution in this solar system is limited very closely to the solar system disk. Pluto is the least massive solar planet.

There are millions of asteroids and thousands of comets in their solar orbits, most of them have small inclinations and are between the orbits of Mars and Jupiter. Many others are orbiting beyond Pluto. Comets, debris and some asteroids have more or less elliptical orbits that can extend from the vicinity of the Sun all the way to Pluto's orbit and beyond (40 or more Astronomical Units). Those pieces can 'fly' near the Earth/Moon on their way to and from their perihelion passages. Some comets hit the Sun/planets/moons, and some comets keep on orbiting the Sun for eons.

Google: **< comets hitting the sun >**

When the deflected objects/asteroids/comets come from the asteroid belt between Mars and Jupiter, they may revisit our neighborhood again in approximately **2 to 15 years.**

Asteroids coming from somewhere beyond Pluto's orbit (40 Astronomical Units) may also have experienced similar deflections of their orbits. They may visit us at any time and then revisit the Sun and Earth at 250 to 400 year intervals if they don't happen to crash into something. If these objects come from 50 Astronomical Unit distance, their orbital periods will be even longer.

22. Living Conditions on Solar Planets and Moons

'O'ur' Sun has nine (Pluto lost planet-status in 2006 but its orbit did not change) major planets and millions of asteroids, comets and other space debris in solar orbits. None of them have an atmosphere humans could breathe. None of them except this Earth has running water, i.e. water in liquid form.

The temperatures on Earth are 'just right' for us. We have good breathing air all around this Earth. **Be grateful to your God/whomever for all of this.** Venus and Mercury are much too hot. One cannot survive for very long without breathing oxygen in kitchen oven temperatures without liquid water, or on the bottomless gas planets, or in temperatures much lower than −40 degrees Celsius or Fahrenheit. There are thousands upon thousands of other things 'just right' here on Earth. All of them cannot be covered in any one book, and many are not even known well.

The following listing of planets shows the impossibility of **any human type life anywhere in this Solar System except on this Earth.**

22.1. Mercury

Mercury is acting as an 'arrow-catcher' for us, as can be seen from photographs of its cratered surface. During the past eons, thousands of comets and asteroids in their orbits around 'our' Sun and other fly-by celestial objects have crashed into Mercury, and many more have crashed into the Sun never to come close to Earth again.

Google: < mercury >
              < the nine planets > Scroll down, and click on Mercury.
http://www.nineplanets.org/mercury.html

The surface on the **dayside of Mercury is hot** at 700 Kelvin = 800 Fahrenheit = 430 Celsius, **and the night-side is cold** at 90 Kelvin = -300 Fahrenheit = -185 Celsius. The temperature dial on most American kitchen ovens goes only to approximately 500 F.

Without a reflecting space suit a 'space-walking' astronaut in a low orbit around planet Mercury would have to contend with the fact that her/his sunny side would have a tendency to heat to a temperature of 700 Kelvin = 800 Fahrenheit = 430 Celsius, and that her/his other side of the body would have a tendency to freeze to a temperature of 90 Kelvin = -300 Fahrenheit = -185 Celsius.

Because Mercury's distance from the Sun is only 0.38 AU (Astronomical Units), its angular distance is never more than approximately 21 degrees from the Sun as seen from the Earth. The intensity of solar radiation on Mercury is about seven times stronger than Sun's radiation here on Earth.

Mercury is not a livable/survivable place.

22.2. VENUS

Venus is also acting as an 'arrow-catcher' for us making life a little safer here on Earth. The atmospheric pressure on the surface of Venus is a crushing 90 times higher than on Earth. Similar pressure prevails in water at 900-meter (3,000 foot = 500 fathom) depth in Earth's oceans, seas and deep lakes. Divers are seldom able to dive to 100-meter (328 feet) water depths, where the pressure is 'only' approximately ten times as great as the normal sea-level atmospheric pressure.

Temperature on the surface of Venus is hot enough to melt lead.

Google: < venus >
              < the nine planets > Click on Venus.
              < jpl solar system dynamics > Under Planets, click on Mean
                                     Orbital Elements.
Venus is not a livable/survivable place.

22.3. EARTH IS THE ONLY KNOWN GOD'S 'GARDEN SPOT' IN THE UNIVERSE

Most of this book is about this Earth. Nothing even resembling life on this Earth has been found (2007) within a distance of eight light-years.

Like the nine (8?) planets and their dozens of moons 'our' **Moon is also acting as an 'arrow-catcher' for us,** making our lives a little safer. Numerous comet/asteroid craters on the Moon are visible to the naked eyes. To see pictures of the Moon, go the Internet site '**The Nine Planets**', and click on the '**Moon**'.

Google: < the nine planets > Click on Earth and then on Moon.

Earth is a very livable/survivable place. For all forms of life, this planet Earth is the nicest place by far anywhere in this solar system. Nothing similar to this Earth exists within at least eight light-years from here. Earth is the only place in this solar system where there is **life supporting air to breathe, water to drink and food to eat.**

Be grateful to your Creator/God/whomever/whatever you think is the 'alpha and omega' of this Earth, 'our' Sun and this Solar System. This Earth could not survive for very long without the life sustaining Sun. Without the craters on 'our' Moon, the other eight (seven 2006) planets and their many moons stopping the incoming asteroids/meteorites/comets, this Earth would be a more dangerous place. "Our' Sun also swallows many asteroids and comets. Some of those objects about ready to dive into the Sun have been photographed.

Google: < latest soho images >

22.4. MARS

Mars is also acting as an 'arrow-catcher' for Earth, making our lives a little safer. The Martian atmosphere is very thin, having an average atmospheric pressure of 7.5 millibars compared to Earth's average air pressure of 1013 millibars = 760 mmHg = 29.92 inHg. The Martian air is not breathable.

If a Mars astronaut on its surface would let his dog outdoors, the dog would die just like many thousands of dogs have died at many animal shelters in their partial vacuum-chambers. **The average Martian surface temperature is 220 Kelvin = -55 Celsius = -70 Fahrenheit.** The range of surface temperatures on Mars can vary from 140 K = -115 C = -225 F to 300 K = +20 C = +70 F.

Mars has two small moons shaped like Idaho potatoes. Even those two little moons have meteor craters.

Google: < **the nine planets** > Click on Mars, and on its moons, Phobos and Deimos.

      < **NASA GISS: Science Briefs: Telling Time on Mars** >

Mars is not a livable place without bringing breathing air/food/heating equipment/water from Earth.

22.5. Jupiter

Jupiter, together with its dozens of moons, is also acting as an 'arrow-catcher' for us, making our lives a little safer here on Earth. Among the myriad of comets crashing into Jupiter was the Comet Shoemaker-Levy 9, which was stopped by Jupiter in July of 1994. For photographs of that event and for more information, type '**comet shoemaker-levy 9**' into **www.google.com** search box.

Google: < **the nine planets** > Click on Jupiter, and on some of its many moons.

      < **comet shoemaker-levy 9** >

http://pluto.jhuapl.edu/soc/

http://pluto.jhuapl.edu/gallery/missionPhotos/pages/032207.html

Jupiter is a gas planet **without having a solid outside surface.** Jupiter has approximately five-dozen discovered moons. Some of the moons have an atmosphere of some sort (maybe very cold methane).

Jupiter is not a livable place and neither are its moons.

22.6. SATURN

Saturn with its dozens of moons is also acting as an 'arrow-catcher' for us, making our lives a little safer here on Earth. Saturn is a gas planet without having a solid outside surface.

Google: < **the nine planets** > Click on Saturn and on some of its many moons.
           < **nasa-cassini-huygens** >
           < **cassini pictures** >
http://www.nasa.gov/mission_pages/cassini/main/index.html

Saturn is not a livable place and neither are its moons.

22.7. URANUS

Uranus is not a livable place and neither are its moons.

Google: < **the nine planets** > Click on Uranus and on some of its many moons.

22.8. NEPTUNE

Neptune is not a livable place and neither are its moons.

Google: < **the nine planets** > Click on Neptune and on some of its moons.

22.9. PLUTO

The International Astronomical Union (IAU) has decided that Pluto is no longer a planet. Of course, it is still the same body orbiting the Sun.

http://www.iau.org/
http://www.earthsky.org/blog/50966/who-else-wants-pluto-to-be-a-planet-again

http://pluto.jhuapl.edu/ This site is about the 'NEW HORIZONS' probe on its
way (2007) to Pluto, arriving there in July 2015.
http://pluto.jhuapl.edu/mission/whereis_nh.php
http://pluto.jhuapl.edu/mission/mission_timeline.html

Pluto is not a livable place.

Google: **< the nine planets >** Click on Pluto and on its moon Charon.
<views of the solar system > Select English, or another language.
Click thumbnail pictures. Click also on People for a list of astronomers.
< mission to Pluto >

There are no more planets than these nine (eight 2006) **planets in this
solar system!**
Extra solar planets are orbiting some unreachable stars more than eight light-
years away.

23. SIZE AND SHAPE OF THE MILKY WAY GALAXY

Recall that with a good pair of unaided eyes, it is possible to see approximately 2,000 stars from one place on Earth, and that all of those stars are Milky Way stars inside a sphere with a radius of approximately two thousand light-years all around us. This visible volume (4,000 light-year diameter sphere) covers only 0.15 % of the total area of the Milky Way Galaxy disk. Recall also that we are about 28,000 light-years from the center of the Milky Way Galaxy, and that the diameter of the Milky Way disk is approximately 100,000 light-years.

Many Milky Way stars, including 'our' Sun broadcast/emit both unhealthy and healthy cosmic radiation sporadically, if not constantly. This radiation is bombarding/irradiating everything around these stars including the planet Earth. The same happens in countless other galaxies also. Fortunately Earth's atmosphere filters the radiation down to healthier levels, and Earth's magnetic field deflects some of the radiation away from us. Here is another good fundamental reason to be grateful to the Creator for our very existence and life.

It is unlikely that the human race could exist under unfiltered/unshielded cosmic radiation. It is also likely that the radiation, which comes through down to Earth's surface, has and has had an important beneficial effect on evolution of earthly species.

The central bulge of the Milky Way is approximately **10,000 light-years thick and the Milky Way pinwheel arm, where we are, is about 5,000 (there are other estimates) light-years thick.** The Milky Way is a pinwheel type spiral disk with several arms containing stars with planets, boulders, etc., dust and gas clouds. Because the Milky Way contains these swarms of stars/pieces of debris and gas clouds, it does not have definite abrupt outside borders. 'Our' **solar system with its nine planets orbiting 'our' Sun is** approximately **20 light-years on the north side of the central plane** of the disk of the galaxy in one of the pinwheel arms.

The **south-end of Earth's spin axis (Antarctica) is pointing toward the galactic central plane**, and the north-end of the spin axis is pointing away from the central disk to approximately 'our' North Star = **Polaris. Polaris happens to be at a distance of** approximately **300 light-years away from us**. For this reason, the density of the star field is generally greater as seen from Earth's Southern Hemisphere than from the Northern Hemisphere.

The spiral arms are spinning around the galactic center, which is believed to contain a massive black hole, or two. **'Our' Sun goes (orbits) once around the galactic center in 250 million years (= galactic year)** at a distance of approximately 28,000 light-years (from the galactic center) with a linear speed of about 250 kilometers (150 miles) per second. Our neighboring stars don't travel in perfect lockstep with 'our' Sun. Some move faster, some slower in their own directions. In addition, the whole Milky Way Galaxy may rotate/roll/spin/wobble as it travels in the universe.

The **shapes of constellations, as seen from Earth, remain basically the same over a human life span. The fastest** (so far found) **'side-ways' moving star is Barnard's star,** changing its position by 10.3 arc-seconds per year or approximately one degree of arc in 350 years. So, for instance, the Big Dipper will look the same to the naked eye for many generations. Most stars have Proper Motions less than one arc-second per year. No wonder fixed stars are called fixed stars!

These **visual** (apparent) **angular motions** of stars with respect to more stable distant galaxies are called **Proper Motions**. As mentioned, the Barnard's star has the greatest known Proper Motion. If Barnard's star would be moving straight toward us or away from us, its Proper Motion would be zero.

Google: < barnard's star >
                 < proper motions of stars >
                 < andromeda galaxy >

Some stars near the galactic center have been 'clocked' going (orbiting) around the galactic center at 5,000 kilometers (3,000 miles) per second. In accordance with Kepler's laws of planetary motions, stars nearer to the galactic center have greater orbital velocities than those stars orbiting at greater distances. The same holds true for solar planets also. For instance, Mercury's orbital speed is 47.8 kilometers (29.8 miles) per second, and Pluto's orbital speed is 4.7 kilometers (2.9 miles) per second.

Google: < table of orbital data for the planets >

The Milky Way Galaxy is similar in many respects to the Andromeda Galaxy. However, Andromeda's diameter is between 150,000 and 250,000 (the latest 2005 number is 220,000) light-years long; and therefore, its area is from twice to six times as large as Milky Way's area.

(Some readers may wonder, "Why about twice as large"? The reason is that the area of a circle is 'Pi times its radius squared'. So if the radii of two circular plates are 'one' and 'one-and-a-half' respectively, then the area of the bigger one is Pi times (1.5 squared), or Pi x 2.25 times larger than the smaller one. Pi is the old constant 3.14159.....)

Most of the beautiful pictures published of galaxies are time exposures, and they have been taken through observatory-telescopes/satellite-telescopes and back-yard telescopes. Observatory and satellite telescopes cost millions/billions of dollars. The 'back-yard telescopes' can also produce images/pictures of objects the naked eyes cannot see and so can regular binoculars. For instance, a few of Jupiter's moons and the ring of Saturn can be seen through good binoculars and through surveying telescopes (theodolites).

A recent (2004) picture taken by the Hubble Space telescope of very distant galaxies was the end result of a total one million second (278 hours) exposure time. The 'cross-hairs' of the Hubble were pointing with high precision to the same spot on the celestial sphere during its many orbits around Earth. It is obvious that during 278 hours, more image producing photons from a very faint object will hit the sensors than for instance during a few minutes of exposure time.

Google: < hubble site >
             < andromeda galaxy >
             < soho >
             < astronomy picture of the day archive > There is a new picture
                for every day dating back hundreds of days. Make an effort to see
                some of these pictures. This Internet site gets much higher than a
                'five star rating'. Pun is intended.

The last Internet site might be well worthy of saving in the 'Favorites'/'Bookmarks' of your computer. The daily pictures are available for about the past ten years. These Internet pictures are newer and more abundant than any printed book can contain. A new picture will be available for the day the reader is reading this paragraph!

24. Water is Essential for Life on Earth

All living organisms on Earth use water. Water **is absolutely essential for all human life**. Water is the necessary key for millions of chemical/physical activities in our bodies. We temporarily own the water and other molecules/atoms/cells while they are in our bodies. Cannibals may have different ideas about this!

Google: < cannibalism in uganda and in congo >

The **human body is the most complicated automatic, physical/chemical, factory, processor and deliverer of its products known to mankind.** On its own, it automatically produces thousands of different kinds of molecules/cells in correct amounts and at correct times as needed. It then automatically delivers them at the right times in the right amounts to the correct places for the daily needs of our bodies. During sleep our bodies automatically rejuvenate and 're-charge our batteries'. Of course, this cannot go on forever due to natural aging. All this requires water in the simple and very complicated processes. **The Creator made it just so!** And then there is much going on that nobody fully knows/understands

The Creator created a masterpiece! Every person is a sample of many marvels. Some of our body parts might be 'limping along' at times. Thank God that the medical profession might be able to restore some of the normal functions when things go awry. Note that the medical profession may be able to restore some body parts toward the God-created normalcy, but nobody has been able to improve on the design. If desired, replace the word 'God' to whatever other individual preference.

As one example of what is happening in our bodies, consider our blood. Our blood cell factories are inside some of our bones, where **blood cells are produced at a rate of** approximately **9,000 million per hour, or 2.5 million blood cells per one second. This could not happen without water.**

It is hard to even imagine that anybody but God could have thought/designed and built such neat, complicated and effective systems as our blood factories and many other systems in our bodies.

Consider our brain, which in some respects operates faster than even the fastest computers (2007). **The brain could not function without water or electrons.**

Google: < where are blood cells manufactured? >
             < complexity of the human brain >
             < red and white blood cells >

Consider our eyes having approximately 125 million of two kinds of receptor cells for incoming light. Those receptor cells convert incident light energy (photons) into signals carried to the brain by electrons along the optic nerve. Our bodies produce at least two kinds of necessary tear molecules for the eyes. The normal human eye lens automatically focuses for near and far objects for a sharp image until around 40 years of age. **Our eyes could not function without water.**

Consider our speaking and hearing. First, there must be air going in and out by the vocal cords. No other solar planet has a suitable atmosphere for breathing and talking. In the fluid-filled inner ear (it is not pure water) there are over 20,000 hair-like nerve cells sending many kinds of sound signals to the brain. The brain is capable of interpreting the qualities of the sound upon reception of the electric nerve impulses. Among the many sounds, we can tell whether a church organ, piano or a violin is playing, or if a child or adult is talking or crying. **Again, without water we could not hear sounds.**

These are just a few examples of the importance of water for the functioning of the human body. All of us received our brain, eyes, ears and other water-based body parts as our birthday present. This all is just impossible anywhere else in this solar system.

Google: < human brain >
             < human eye >
             < human inner ear >

Of course, we want to have, and most of us do have, water 'on tap' for many purposes in daily lives. The Sun supplies continuously fresh water by evaporating every year approximately 1.25 meter (4 feet) thick layer off the oceans to the atmosphere. Of course, the Creator even supplied the watery oceans.

The mass of the annually evaporated water amount is approximately 109,400 cubic miles (456,000 cubic kilometers), or 4.6 x 10E14 metric tons.

The daily evaporated amounts of water are: 300 cubic miles = 1.250 cubic kilometers = 1.26 x 10E12 metric tons.

The hourly evaporated amounts of water are: 12.5 cubic miles = 52 cubic kilometers = 5.25 x 10E10 metric tons.

The same rate prorated per one second is: 14,600,000 metric tons per second, 24 hours a day, 365.242 days a year, year after year for eons. Recall that **71 % of the Earth's surface area is ocean**, and the sunlit area on the average is always 50 % of Earth's surface area day and night.

The evaporated water forms clouds, which carry the evaporated water over most places on Earth. It eventually rains back to the oceans as well as on land areas. The rainwater keeps up the lakes and reservoirs, the rivers flowing and the water table in the ground. The average annual rainfall on all land areas is 90 centimeters = 35 inches. On some Hawaiian islands, the average annual rainfall is much greater; in the Sahara desert it is much less. The mass of the average annual rainfall is 890,000 metric tons per square kilometer, or 2,300,000 metric tons per square mile. The continents receive precipitation each year equivalent to a layer of water 73 centimeters (29 inches) thick over their entire surface areas, of which 42 centimeters (17 inches) is lost to evaporation and 31 centimeters (12 inches) make up the runoff.

Without liquid water there would be no agriculture, trees nor food to eat. The barren Moon, planet Mars and some deserts on Earth are good examples showing what happens when liquid water is missing.

There are theories about how and from where the Earth's surface got its water. They are not covered in this book. Volcanoes are one source in water production.

Google: < composition of volcanic gases >

25. Search for Life Elsewhere in the Universe

So, this Earth is the 'one and only' place in this Solar System capable of supporting human-type life. It is well known that no human or alien-type life can exist on any other solar planet or on any of their moons. One can safely ignore all stories about any Martians or other extra-terrestrial Aliens ever having been here on Earth. Their possible appearance here on Earth is totally impossible for several good reasons.

If a question comes to mind 'Where is Heaven?' and 'Where is Hell?', the answer cannot be found in this book. Heaven and Hell are not on any mountaintop or anywhere under the ground inside this Earth. Among the possibilities, consider them to be spiritual quantities or consider them to be in another dimension, accepting that all human knowledge is incapable to produce definite answers about the physical locations for Heaven and Hell. There are many other 'things' we simply just do not know.

The hunt for existence of life elsewhere in the universe generally requires the availability of water between its freezing (ice) and boiling points (steam) on the prospective celestial body. So far, no form of higher human-type life has been found anywhere else but on this planet Earth. There may be civilizations elsewhere, but they are so far away that the traveling times either way would be many centuries long, and one-way radio contacts will take more than ten years. It will be a slow conversation even after we learn to understand the languages in the incoming signals

The whole human life is a great admirable wonder and a miracle in a million ways. We did not – repeat: we did not, nor did any of our ancestors do much of anything about our creation, evolution or creating the Sun or Earth. They are available here for our use. We are only the users/developers/stewards of some acreage and some physical things/items only for our life times.

Think about your possessions/acreages/items. Where were they 500 years ago? Who owned them 500 years ago, and who will take care of them 500 years from now? We just don't know.

God/Creator, or something else was required to provide it all here on Earth. **If you value** (you do, don't you!) **your existence and life here, PRAISE YOUR LORD, or PRAISE whomever** did it all according to differing individual preferences! As a gift from God/Whomever, we have our bodies and the unique opportunity to live our years in our bodies on this wonderful planet Earth.

Of course, it is possible to develop better tomatoes, apples, corn and many new agricultural/animal/food-pyramid products from those that already exist here on Earth. However, all these developments have started from those, which God/whoever originally has put here on the surface and into the ground of this planet for our use/study/learning/cultivation/disposal and development. God/whoever practically put his gifts in our laps. He/Whoever included minerals/metals/oil/gas/trees/crops/etc. for us humans to find/eat/drink/develop and use. **He/whoever gave us lots of good work to do! We should be greatly thankful for it all!**

SETI Institute conducts scientific research on Life in the Universe with an emphasis on **SETI, the Search for Extraterrestrial Intelligence.** The search goes on. Over 200 extra solar planets with no life have been found so far (2007).

Google: < seti >

25.1. POSSIBLE LIFE ON INNER AND OUTER SOLAR PLANETS

Both inner solar planets (Mercury and Venus) **are too hot and all six** (five, 2006) **outer planets and their moons are too cold for human type life. In addition, their atmospheres (if any) are not breathable.**

There is absolutely no hope of finding significant life on any other solar planet or on any solar system moon/asteroid/comet. This is contrary to the beliefs of a few suicidal cult members, who believed that after committing their suicides in 1997, they, or their souls would go on Hale-Bopp comet/Heaven to live happily ever after. If those cult members got their wishes, even their souls (if they still have them) are now frozen down to temperatures very close to the absolute zero. When the comet comes back for its next perihelion visit in 2031, the poor souls (if there) will be warmed by the Sun – by how much, is unknown.

It is an undeniable scientific fact that in this Solar System, except on this planet Earth, there is no other planet/moon/asteroid/comet/object or anything else suitable for human-type life. We are lucky to be here. **Be grateful to your Creator, or to whom you think is responsible for your good luck!**

Google: < comet hale-bopp >
         < recent comets >
         < glossary astronomical unit {AU} >
         < technical facts of planet mars >

Even planet Mars at 1.52 AU (228,000,000 kilometer = 142,000,000 miles) average distance from the Sun is too cold for human comfort. The average temperature on Mars is –81 F (-63 C, -210 K). The Sun's radiation intensity on Mars is only 43 % as strong as the radiation Earth receives. Earth's distance from the Sun is just right.

From the Quaoar's distance at 42 Astronomical Units, 'our' Sun looks as small as one-quarter coin ($0.25) at a distance of 0.9 miles (1.5 kilometers). The angle extended by that quarter coin is 3.3 arc seconds, and also the angle in which 3.3 millimeters (0.13 inches) can be seen at a distance of 206 meters (675 feet). To really see anything so small that far away, a good telescope under stable atmospheric conditions is required.

For comparison, from Earth, the visual angular diameters of both the Sun and Moon are approximately 0.5 degrees, or 1,800 seconds of arc.

By the inverse square law, the Sun's radiation at 42 Astronomical Unit distance is only 0.06 percent as intense as it is here on Earth. This shows how very small amounts of energy solar panels at 42 AU (Astronomical Unit) distance could collect, and how cold it must be 'over there' without take-along heaters and insulation.

For comparison, solar panels parked on Earth's surface at mid-latitudes can collect during midnights from the Milky Way stars almost the same tiny amount of energy as they could collect from the Sun when being at 42 Astronomical Units from the Sun. The solar panels on satellites in near Earth orbits can collect more than 1,750 times more energy from 'our' Sun than they could if they are taken to a distance of 42 Astronomical Units.

Google: < inverse square law >

The outer planets Jupiter and Saturn are gas planets without a solid outer surface to stand on. There can be no human-type life on planets Uranus, Neptune, Pluto or on any of their moons. They are nice comet/object catchers for us on Earth.

Google: < planet jupiter photo gallery >
               < planet saturn >
               < planet uranus >
               < planet neptune >
               < planet pluto >

25.2. REASONS WHY OTHER SOLAR PLANETS/MOONS ARE LIFELESS

1.   Environmental Conditions

Numerous environmental/living conditions on all other eight solar planets and on their moons are such that all continuous, or **even a 'quick' visit to most of them would be of very short duration or even deadly.**

2.   Atmospheres

Atmospheres 'over there' are not fit for breathing. To survive we all need to inhale air/oxygen molecules several times every minute. Possible astronauts 'over there' can talk with each other by using two-way radios (and maybe sign-language), but they cannot step 'outdoors' into the 'too thin/dense/hot' or poisonous atmosphere and live without carrying their own breathing air/oxygen and cooling/heating equipment. Without oxygen any person will choke to death in a very short time.

If one is not capable to inhale and exhale, voice communications 'over there' will be impossible. Sound from the vocal cords of one astronaut would not carry to the ears of another astronaut. Vocal cords cannot produce words/sounds without inhaling/exhaling oxygenated, preferably moist air/oxygen. Breathing pure oxygen for extended periods of time is not a very good idea. Cell phones would not work 'over there' either, because there are no cell phone towers and no other necessary associated equipment.

Similar impossible situations exist also for all living earthly animals if they are taken 'over there'. It can be **safely said that in 'our' Solar System, this Earth is the only place where the birds sing and where green grass, plants and trees grow**. Earth is also the only place in 'our' Solar System, where one can smell roses unless they are delivered from Earth. However, finding bacteria or some such lower forms of life on Mars, or on some of the Jupiter's moons, is possible.

Astronauts can live in their heated/cooled space suits, and/or in air tight space ships carrying their own breathing air as they already have done on the Moon. Similar astronaut visits may be remotely possible to planet Mars and possibly to some of Jupiter's moons. Such visits will be exceedingly difficult and dangerous.

3. Great Distances

One very important reason for the difficulties and limitations for space travels are the great distances. The distance from Earth to Jupiter and its moons ranges annually from 6.2 to 4.2 AU (Astronomical Units), which is approximately 2,600 to 1,750 times as far as the Moon is from Earth. **Even a one-way radio message to/from Jupiter will take from 35 minutes to 52 minutes, depending whether Earth and Jupiter are on the same side or on the opposite sides of the Sun.**

The distance to Mars is 'only' from 0.4 to 2.7 AU, meaning that one-way radio communication will take from 200 to 1350 seconds (3 to 22 minutes).

Google: < overview of the solar system >

4. Temperatures and Pressures

Temperatures 'over there' on other solar planets and on their moons are either too high or too low for water to be between its freezing and boiling points. For instance, one measured unbearable temperature on the surface of planet Venus was 870 degrees Fahrenheit = 465 degrees Celsius = 737 K. Capital K stands for Kelvin degrees.

Google: < kelvin temperature scale >

Also, the atmospheric pressures 'over there' are not friendly for visiting astronauts. The bone-crushing atmospheric pressure on Venus is 90 times greater than here on Earth. Vacuum conditions on the Moon and the low atmospheric pressure conditions on planet Mars require pressurized suits and pressurized space ships for the duration of the stay.

A pressure of 90 times Earth's sea level atmospheric pressure can be experienced in water (on Earth) at depths of 900 meters (2950 feet). This pressure is ten times greater than what divers (in water) can endure for short periods of time. Venus is truly a pressure cooker for anything on its surface!

On planet Mars the surface temperature was once measured to be – 82 degrees Fahrenheit = – 65 Celsius = 210 Kelvin degrees, which is much too cold for comfort. In addition, the thin Martian atmosphere is not breathable. The Martian windstorms move the dust at high speeds up to 50 meters per second (100 miles per hour). Because the atmosphere is so thin on Mars, a standing person would not necessarily get blown over by the Martian 100 MPH winds, but she/he and the eye shields of the helmet would get sandblasted pretty well.

5. Cosmic Radiation

Cosmic radiation is more dangerous on Mars than on Earth, because Earth's atmosphere and magnetic field reduce its intensity to acceptable levels on Earth's surface. Radiation is very strong near Jupiter. Shielding space travelers from cosmic radiation is also a problem.

6. Gravitation/Gravity

For us Earthlings to survive, gravity must not be far from 1 G (1 G= 9.81 meters per seconds squared = 32.2 feet per seconds squared) for many good reasons. It is the 1 G gravity that gives our body mass its weight, which bathroom scales indicate. We weigh a little more near the poles than near the equator, as well as at low elevations than on high mountaintops. Some recently (2007) found extra-solar planets have masses (and therefore gravities) over ten times Earth's mass/gravity. Those distant solar systems can also have smaller Earth-like planets but so far nobody knows much about them.

Ten G gravity is deadly for humans. Such gravities would flatten us permanently against the ground or against the outer wall of a centrifuge.

Even a continuous 2 G gravity becomes very uncomfortable, which could be demonstrated in a centrifuge by producing 1.73 G horizontal centrifugal acceleration, which combines with the prevailing 1 G vertical gravity resulting in 2

G gravity perpendicular against the tilted outer wall of the centrifuge when the upper part of the wall is tilted out by approximately 30 degrees from the vertical. A person could live there, test herself/himself and walk around that tilted outer wall under the influence of 2 G gravity for as long as desired.

The outer wall could be permanently built to the 30-degree tilt for 2 G gravity, or the wall panels could look somewhat like the sliding panels on an incoming airport luggage-carousel, on which one could walk around the tilted outside wall of the centrifuge under desired variable gravities without slipping/sliding down on the tilted outer wall of the centrifuge. A carpenter's level placed on the tilted walkable conic circular wall (= floor/horizontal surface for the person in the centrifuge) surface would indicate the wall to be a level surface when the tilt and the spinning speed are adjusted to desired values. If the occupant held a plumb bob hanging from her/his finger, the plumb line would be perpendicular to the tilted walkway. The plumb bob would also feel twice as heavy as normal.

For a 3 G gravity pulling/pushing the occupant against the tilted centrifuge wall, the outer wall should be tilted out by 19.4 degrees from the vertical.

For a five G gravity perpendicular against the tilted wall, the outer wall should be tilted by 11.5 degrees from the vertical.

For ten G gravity perpendicular against the tilted wall, the outer wall should be tilted by 5.7 degrees from the vertical. **For 10 G gravity, the centrifuge becomes an execution chamber in a minute, or in just a few seconds.** See what you would weigh: 'Your Weight On Other Worlds' at:

Google: < exploratorium.edu/ronh/weight/index.html >

Merry-go-rounds in amusement parks are centrifuges, and so are the household washing machines when in their spin cycles. The typical spin rate of a washing machine tub is from 500 to 1,600 RPM. If the radius of the spinning tub is 25 centimeters (10 inches), the centrifugal acceleration at the tub-wall ranges from 70 G to 700 G. If some Dennis-the-Menace-type little boy forgot a dead mouse into his dirty pants pocket, the mouse will come out from the washer looking like it was run over by a truck.

It is unlikely that anyone would like to spend much time under the influence of that 2 G or greater gravities. However, that would be 'something new' for the Guinness Book of Records.

Under the influence of 2 G gravity we all would weigh twice as much as we do now. Even dieting cannot change that!

Our walking/running would be like carrying an identical twin on our backs.

Lifting a 40-pound bag of material (maybe salt) would be as hard as lifting an 80-pound bag right now.

Our car with a 200-horsepower engine would be as sluggish as the same car with a 100-horsepower engine.

If a car is accelerating at a uniform acceleration from a standstill to 60 MPH = 96 kilometers per hour in four seconds, the average horizontal acceleration pushing the driver against the seatback is approximately 0.68 G. During those 4 seconds, the driver is under a total resultant acceleration of 1.2 G, which is obtained as a vector sum of the horizontal 0.68 G and the prevailing vertical 1 G acceleration. If the driver's normal weight is 150 pounds = 68 kilograms, during that acceleration he weighs 181 pounds = 82 kilograms. This could be measured by weighing the driver during the four-second acceleration if she/he sits on a properly installed bathroom scale.

Present airplanes would be twice as heavy as they are right now.

If gravity on Earth would increase to 2 G, the atmospheric pressure would double assuming that the mass of the atmosphere would not change. The atmospheric pressure on our bodies would be approximately the same as it is at ten-meter (33 feet) deep water.

Recall that 'our' Moon's gravity 0.16 G (= 160 centimeters per seconds squared = 160 gals = 160,000 milligals) is approximately **one sixth of Earth's gravity.** When the gravity is that low, **any celestial object,** with the kind of sunshine the Earth and Moon (and asteroids and comets in our neighborhood) receive, **cannot keep their atmospheric gases from escaping into the vacuum of space.** Therefore, small solar system asteroids/comets/moons don't have an atmosphere. One cannot breathe in a vacuum.

Solar radiation pressure pushes masses away from the Sun. For instance, loose small dust particles on and near a comet form a tail for it because comet's own gravity is weaker than the push of the solar radiation pressure at the distances of the small particles from the comet's nucleus. When a comet gets far enough from the Sun, the solar radiation pressure will get weaker than the comets own gravitational pull on the small dust tail's particles. Then some of them will settle back to the comet's nucleus, **and after that, the comet becomes an asteroid!** These processes are repeated at the comet's next perihelion passage.

Google: < tse-pageos >
        < solar sail info >
        < solar electromagnetic radiation pressure index >

If some of the comet's tail particles got 'too far' away, they will not settle back to the comet's nucleu, but they will continue to orbit essentially in the comet's orbit behind the nucleus causing the sand-blasting meteor showers. The Earth can zip through that ancient meteor tail, which can be a scattered wispy cloud extending around the whole comet's orbit like an elliptical reef, or like a smoke ring from cigarette smoke. One example of this is provided by the annual Leonid left-over tail particle meteor shower from comet Temple – Tuttle crossing Earth's orbit every day around an area where Earth flies through it at 30 kilometers (20 miles) per second in the latter part of the month of November. The Leonid particles themselves are incoming to Earth almost on a head-on collision course at speeds up to 40 kilometers (25 miles) per second. In 1966 the maximum measured rate of Leonid meteorites was 500.000 per hour, and the maximum measured incoming speed was 71 kilometers (44 miles) per second (71= 30 + 41 kilometers per second).

Google;   < meteors and meteor showers > Scroll to see the orbit of debris.
        < annual meteor showers > This site lists dozens of annual meteor showers.

If small asteroids would have loose powdery stuff on their surfaces, they would be called comets when they come sufficiently close to 'our' Sun showing their debris tails. The tails point away from the Sun due to solar radiation pressure.

'Our' Moon has loose powdery stuff (regolith) on its surface, but the Moon's gravity (at Moon's surface) is stronger than Sun's radiation pressure. The solar radiation pressure is not strong enough to blow the Moon dust away and therefore, 'our' Moon does not have a tail! If it did, wouldn't that be something to see!

Google: < **lunar regolith** >
               < **escape velocity:planetary reckoning** >

Some astronauts have spent a few months in near zero gravity conditions. Under those conditions human bones and muscles suffer and the general human health could not be maintained for many generations. Centuries would be required to reach any 'unreachable star' with its solar system and Aliens.

It is obvious that 1 G is just about right for all of us.

It is also obvious that a livable planet must have an oxygenated atmosphere with a decent pressure. Being in a vacuum brings almost an instant death to a person who is not wearing a million dollar airtight space suit.

Google: < **the nine planets** >
               < **martian atmosphere** >
               < **martian dust storms** >

That is basically 'it' for finding intelligent living beings outside of this Earth in this Solar System. Jupiter and Saturn themselves are so-called Gas Giants without a solid outer surface. Mercury's airless surface is way too hot on its sunny side and way too cold on its night side. The outer planets Neptune, Uranus and Pluto are too cold and too far for astronaut visits in the fore-seeable future.

The possibility of extra terrestrial life is under continuous study. There are researchers/scientists who devote much of their time for searching for extra terrestrial life.

Google: < **seti institute** >

This Earth is the only place for all of us to live our life times in our bodies. **This is a unique planet. There is nothing else like it within many light years.** We value our lives, existence and the large number of favorable possibilities here on Earth. We should be extremely grateful to our Creator/whomever that we were born to live here on this wonderful planet Earth. None of us are here by our own designs/plans. It took a while even for our parents to know if their child (= you and I) was going to be a girl or a boy.

This Earth was made for us with extreme skill and consideration of many variables, which only the Infinite Supreme Being could do. Nobody fully understands it all.

Recall that the many spiritual/religious possibilities/considerations are out of the scope of this book.

26. We Will Never Meet Aliens from Extra Solar Planets

Journeys between us and any other solar system many light-years away possibly having livable planets would last several centuries. The traveling would have to go through dangerous cosmic radiation 'environments'. The required tremendous amount of energy to lift an inter-stellar space ship off its launching pad, and all the energy required to propel it to sufficiently high speeds, rules out all possibilities for successful interstellar travel from Earth. And still, a large amount of energy is needed for safe braking/landing at the destination.

Imaginary aliens traveling to this Earth would have to overcome similar 'travel restrictions' as humans would face in any similar inter-stellar travel. The restrictive conditions are overwhelmingly insurmountable for all mankind. Some of the reasons are detailed below.

1.   The simple answer is 'NOT POSSIBLE' to all thinkable transports of human type life over inter-stellar distances. There are several valid scientific and practical reasons for it.

So far, nobody even knows where other *life supporting* planets might be orbiting their stars (stars to us, suns to them). Therefore, we don't even know in which direction should one travel. There are some good possibilities for life supporting planets around some stars, and there are also stars, which don't qualify for several reasons.

Recall from earlier that the search for extra-solar planets is ongoing. It is known that planets beyond this solar system (extra solar planets) are at least eight light years away from us, meaning that a **two-way radio communication over that distance would take a minimum of 16 years. The travel time would be several centuries by a space ship.** Long trips lasting centuries in one spaceship are impossible for human beings and for incoming Aliens.

The nearest star to us, Proxima Centauri, is 4.24 light-years away, but it cannot have livable planets reasonably close to it for reasonable life-supporting temperatures. One reason is that many stars are doublets (double stars). If there would be a planet orbiting a doublet at a livable distance (warm enough), the planet would eventually/soon crash into one of the doublet stars because a three-body system is an unstable system for long time periods. If the planet is further out for a safe orbit, it is too cold for human type life.

Google: < nearest 50 stars >
               < planetary three-body problem >
               < nearest doublet stars >

2. **At a highly-optimistic imaginary speed of one percent the speed of light**, or 3,000 kilometers (2000 miles) per second, one-way travel time over an 8 light-year distance would be 800 years. It would be very difficult and beyond the present technology to propel a heavy space ship to such a high speed. The required time of acceleration to reach a velocity of 3000 kilometers (2000) miles per second at one G acceleration would be 83 hours. At an acceleration of 2G, the required time would be 42 hours. That would be a long rocket burn of lots of fuel and oxygen, or nuclear fuel until the weightlessness 800-year travel could start!

3. Consider the **minimum energy required** for lift-off from the launching pad and then to propel a Space Shuttle size space ship and onboard fuel/food/equipment to the speed of one percent of the speed of light. That velocity is 430 times greater than the Space Shuttle velocity of 7 kilometers (4.4 miles) per second in its normal orbit around Earth. To find out the required amount of energy, 430 must be squared obtaining 184,000 times more energy than the present Shuttles carry without considering the required extra weight.

Google: < kinetic energy >

Therefore, the energy required to propel the Shuttle size space ship to a speed of **just one percent of the speed of light** requires 184,000 times more energy than what the present Shuttle fuel tanks hold! The traditional fuel tanks of the Shuttle have been shown on TV many times. Multiply that by 184,000, and that is not even all that is needed! **The masses of the extra fuel and the bigger fuel tanks** also require even more lift-off energy; still more energy is needed to propel 'the whole thing' to the speed of one percent of velocity of light or more.

4.   **Another large amount of energy** would be required for a safe landing when approaching the destination planet. The reason is: **at the end of the journey (more than a minimum of eight centuries later), the speed of the spaceship must be braked down for a safe landing**. After 800 years the fuel needed for braking might have 'gone sour'. Atmospheric friction at the end of the journey cannot do the required braking for a safe landing.

5.   In addition, **the housekeeping energy on the spaceship** would be required during most of the journey because solar panels are useless in trying to capture enough radiation energy from starlight when traveling at great distances from the nearest stars, one of them being 'our' Sun. One can compare this to the amount of energy solar panels parked on Earth can collect on winter midnights from the clear moonless sky.

6.   Even **using nuclear power equipment,** the energy requirements for lift off, acceleration, house keeping and braking for the destination would be much greater than anything possible at the present time

7.   Still another great problem for the space travelers is the **dangerous cosmic radiation in space coming from many directions.** For instance, when the Galileo spacecraft flew past Jupiter's moon, Amalthea, in November 2002, it was exposed to radiation more than 100 times a lethal human radiation dose. Shielding for the cosmic radiation would add to the weight and energy requirements, and probably no reasonable shielding would be adequate for radiation exposures lasting for centuries.

    This Earth, this solar system and the volumes of space beyond are constantly bombarded by lethal radiation and by many elementary particles. The intensity of this radiation and particle bombardments is reduced here on Earth's surface by the filtering effect of Earth's atmosphere and also by partial deflection by Earth's magnetic field. A space ship would require extra shielding adding to its mass.

Google: **< cosmic radiation in our galaxy >**
                   < NASA/Marshall Solar Physics > Scroll down to Recent Solar
                   Physics News Stories.

Filtering and deflection effects of dangerous radiation away from us are another of the many beneficial functions 'our' atmosphere and magnetic field have
in making our lives possible here on Earth. This is another fact demonstrating
the infinite wisdom, ingenuity and capability of the Creator! We are very lucky to
be living on this wonderful planet Earth, which is in marvelous balance in more
ways than anyone of us knows.

8.   Still, another problem for space travelers is the **fast moving asteroids and
other pieces of rock and ice (called meteorites after they enter Earth's atmosphere)** that have nicked some windows of the Space Shuttles. Those pieces can
have relative speeds with Shuttles between of 12 and 72 kilometers (7.5 and 45
miles) per second here at one AU (Astronomical Unit) distance from the Sun.
These speeds are for those pieces of mass, which are traveling around 'our' Sun
in more or less eccentric elliptical orbits. Velocities of interstellar pieces can be
much higher.

**Collision velocities in inter-stellar space can be much greater with
the fast traveling space ship**. Head-on collisions occur at higher relative
speeds than hits from 'behind'. **Collisions with a fast moving space ship,
such as the mentioned 3000 kilometers (2000 miles) per second, could
easily be lethal.**

Asteroids/meteorites can have **masses from a pebble of gravel to a bowling
ball to many tons.** Most collisions with large objects would likely be lethal to
space ship travelers.

The Peekskill Meteorite Car was hit by a bowling-ball-size meteorite at a
'slow' speed (due to atmospheric braking) of approximately 100 meters (300 feet)
per second. When it entered Earth's atmosphere its velocity was between 12 to
72 km/s (8 to 45 miles/s). Imagine what would have happened if the velocity had
been 300 times greater (30 km/s = 19 mi/s) without atmospheric braking when
the kinetic energy would have been 90,000 times greater! That kinetic energy was
spent to heat the air along the meteorites path to Peekskill.

One could say with 'tongue in cheek' that the **incoming meteorites contribute to global warming!** They do but only by a small amount.

By some estimates, the accreting meteorite mass floating down upon this Earth is approximately 78,000 tons/yr, or 214 tons/day. It would be rather easy to compute, how much heat energy 214 tons of meteorite pieces produce per day when their kinetic energy is converted to heat from their average incoming speed of 40 km/s (25 mi/s) to 100 m/s = 330 feet/s. This heat energy is in addition to Sun's regular daily heat radiation.

Some fast moving incoming comets into this solar system are deflected more or less to hyperbolic solar orbits by some solar system masses sending them out of this solar system if they don't collide with anything substantial. They are then just one-time visitors. Their speeds can easily be 300 km/s (200 miles/s), or more with kinetic energies 9,000,000 times greater (or more) per a unit mass than the Peekskill meteorite had before entering Earth's atmosphere.

Google: **< peekskill meteorite car >**
              < kinetic energy >
              < terrestrial impact craters >
              < 10-Accretion of Mass >

Those Aliens, who would be traveling to Earth from a distance of at least 8 light years, would face similar traveling problems as described above for travelers from Earth.

It is safe to say for the above-mentioned reasons, no living Aliens will make it to Earth from any other solar system.

Earthlings simply cannot travel to any *extra* solar planet.

There are no possible living places in 'our' Solar System for any Aliens.

In summary, no one has ever, or will ever meet any living Aliens on this planet Earth *from any other planet*.

27. KEPLERIAN ORBITS

For quicker understanding, it is beneficial to know the **meaning of a few words used to describe Keplerian orbits** (Johannes Kepler (1571-1630). Orbit determinations require knowledge/mathematics/capable computers and education in Celestial Mechanics. Mathematical details are not discussed in this book.

Google: < kepler's laws with animation >
               < planetary orbital elements >
               < satellite observing: orbital elements >
               < physics-celestial mechanics >

The central body is at one focus of the orbit ellipse. **This is Kepler's First Law.**

If a body is in a **circular orbit**, the two foci of the orbit ellipse have come together to the center of the circle. Due to gravitational pulls of many 'nearby' objects, solar radiation pressure, etc., a circular orbit does not stay perfectly circular for very long.

If the orbit of an object is a **parabola** or a **hyperbola**, it is in a fly-by situation; such a fast mover is only deflected by the Sun/planets into a new direction/orbit. Such an object will never come back. Parabolas and hyperbolas are theoretical possibilities.

Google: < parabola >
               < hyperbola >

All solar and Earth orbits of interest are ellipses. Orbit ellipses may and do change in shape and size into new ellipses for several reasons. **Celestial Mechanics** is the branch of physics/astronomy, which deals with orbits and computes them for desired instants. A satellite may be orbiting 'our' Sun, Earth,

Moon, Mars or some other object. Only a few items of Celestial Mechanics are even mentioned in this book.

Google: **< foci of an ellipse >**

Perihelion = pericenter is the nearest orbital point to the Sun/central body of an object orbiting the Sun/central body (Helios = Sun).

Aphelion/apocenter is the most distant orbital point to the Sun/central body.

Perigee is the nearest orbital point to Earth of an object orbiting Earth (Geo refers to Earth).

Apogee is the most distant orbital point to Earth of an object orbiting Earth.

Eccentricity of an orbit ellipse describes how flat or how round the orbit is. It is a constant number between zero and one. Eccentricity of a circle is zero. Eccentricity of Earth's solar orbit is 0.0167. Eccentricity of a parabolic orbit is one. Eccentricity of a hyperbolic orbit is greater than one. Eccentricity of a straight line is infinitely large.

Google: **< conic sections >**

The following numbers are a sample of comets' eccentricities from a table in **Google: < asteroids and comets >**: 0.540, 0.383, 0.623, 0.995, 0.9998 and 1.000019 for McNaught comet. When the eccentricities are almost one, the semi-major axes of the orbits are long. At their perihelions they travel at their fastest speeds swinging quickly around the Sun. After these perihelion passages, they climb up (from the Sun) toward their distant aphelions constantly slowing down to their minimum speeds at their aphelions. From aphelion they start falling back (toward the Sun) with constantly increasing speeds to their maximum speeds at their next perihelions repeating their previous orbits, although the next orbits might be perturbed (changed) by some of the solar planets/moons/objects along 'the way'. For hyperbolic orbits like the one for comet McNaught with its eccentricity 1.0000019, the aphelion is at infinity and such a comet will never 'fall' back toward the Sun. In due time, astronomers will publish new findings from their observations, measurements and computations for the next asteroid orbits.

Google: < **ellipse calculator** >
               < **orbits of comets** >
               < **planetary orbits** >
               < **navy and satellites** > Click on 'orbits'.
               < **asteroids and comets** > This site has more than a dozen informa-
                                       tive links.

28. Original Building Blocks of all Materials

Atoms/subatomic particles/molecules/cells make up all material items in this solar system including the whole planet Earth and also all human bodies. It is believed that the atoms we know were created in star explosions a few billion years ago. These particles did not come into being by themselves from nothing. Whoever/whatever created/made-it-so is beyond human detailed knowledge and understanding. We know that the material exists, and some of it is even alive – like our bodies! Information on all existing elements is listed in the periodic table.

Google: < periodic table of elements >

About one hundred different types of atoms form millions of various types of molecules forming all material things/objects/liquids/gases/cells here on Earth. **Atoms/molecules form the cells/mass of our muscles, bones, skin, internal organs and every item within our bodies from head to toe. All of us live our lives here on Earth in our bodies. All of this involves more miracles than any one can ever know or explain.** Of course, much is also known, and the knowledge is increasing every week.

We don't eternally own the molecules, which are presently in our bodies. We just have them on loan from Earth/Sun/Solar System and Milky Way resources. Most of our molecules/atoms/cells came from earthly materials. Some molecules/cells of our bodies are manufactured with the help of the Sun's radiation.

For example, the ultraviolet-B radiation from the Sun hitting our skin is responsible for producing Vitamin-D (Vitamin-D is a hormone) for our general well being. 'Too much' of the Sun's radiation can be deadly. Moderation is the 'word', and on Earth, we have many choices.

Google: < sun's radiation >

             < vitamin d >

Some molecules will stay in our bodies longer than some other molecules. For instance, we may drink from one to two liters (one quart to half gallon) of water every day. Annually this amount of water can add up to half a ton or one half of a cubic yard (half of cubic meter). Water is continuously used in/by our bodies for thousands of chemical/physical processes. An average adult human body is 55% to 75% water.

28.1. ORIGINS OF ATOMS AND MOLECULES

Atoms/molecules/subatomic particles present on/in Earth and in the solar system/universe have existed in some form for a long time. **The visible universe is presently estimated to be 13.7 billion (13,700,000,000) years old.** Was everything created at once in the Big Bang, or is the creation an ongoing process? Wish we knew for sure. Atomic particles, atoms, molecules and radiation are created almost continuously in star explosions taking place in many galaxies including the Milky Way Galaxy. They can be the building blocks for new solar systems far in the distant future. Our Milky Way Galaxy produces approximately 7 new stars per year and approximately two supernovae per century (2006).

Google: < age of the universe >

             < big bang theory >

             < NOVA online, runaway universe >

Whoever or whatever the Infinite Creator/God is, it is clear that all persons, the whole planet Earth and this solar system are a part of the creation/evolution composed of atoms and molecules, which in some way came together to form this solar system and this Earth. And miraculously, life on Earth started in some way.

28.2. EVOLUTION/CREATION, OWNERSHIP OF THE PLANET EARTH

The Creator (through the Big Bang?) created even the evolution. If not, where did evolution come from by itself, and how could it exist, continue and be possible without having some kind of starting point (creation) and association with existing atoms/molecules/cells? **If God didn't make it so, then what, or who did?** No one knows for sure. **How else would evolution be possible if situations and places were not created where evolution is possible? Each person is entitled to have her/his own opinion on these very deep eternal questions, which have not been completely explained to everyone's satisfaction.**

We do know that we exist and that this solar system exists and that we cannot nor could our ancestors have created anything similar.

No one can even tweak the Sun's intensity of radiation up or down in any way. Even if humans could adjust 'our' Sun in some way, the end results/consequences probably would be disastrous for us on Earth. Humans just don't have the necessary knowledge/farsightedness for it. So, we better be deeply thankful and appreciative to our Creator/Whomever, who made it 'just so'!

Google: < the solar and heliospheric observatory > This Internet site is well worth entering into Favorites/Bookmarks of your computer.

How about the ownership? Who owns 'our' Sun, Earth and the other planets? **We, some families and the states can have legal ownership rights to some acres (hectares) of some land areas here and there on the surface of this Earth for some years, decades or maybe centuries.** Who owned that land five hundred, or 10,000 years ago? Who will own it five hundred years from now?

The original and final ownership of this Earth and Sun belongs to God/Creator/Supreme Being or whatever one wants to call it. Humans can only have a temporary stewardship-type ownership of any land on Earth. God/Creator/Supreme Being allows us to live here on His property. He does not collect any rent! We have a pretty good deal here. We are just very temporary stewards and small caretakers. We can make some small local improvements here and there. We can also mess up some things, which we naturally want to avoid.

28.3. DID GOD DO IT ALL? WHAT IS CONTROLLING OUR SUN?

In addition to all material things, there is the miracle of life in living humans/animals/insects/plants/fungus and in all living things on Earth.

None of us made the decision when or where on Earth we were going to be born. But it did happen. Did God/Creator have anything to do with it?

So far, no one has been able to create one living thing starting even from those atoms/molecules the Creator put here on Earth for our disposal. So far (2007) humans have not created any new bacteria without starting from mutations and starting from something other materials than what God/Creator/Whoever already had created and put on our disposal.

However, it would not be surprising if some scientists in the near future can create some type of life; for instance, some bacteria to begin with. That kind of development will likely bring Nobel Prizes. Many living things have mutated on their own in the situations they happened to be in. Some mutations are also made by human intervening actions. **But even mutations of living cells cannot start from non-living molecules/atoms – at least until this time. Existing seeds of life – at least so far – are required for all mutations.**

Not even one blade of grass is man-made without mutating or starting from some already living/existing plant/cell, or seed. **The origins are from God/Creator/ Whomever** in ways no human being fully understands. **God the Creator, the Supreme Being, is the alpha and the omega, the beginning and finally the end, when 'our' Sun becomes a red giant star, and all life on this Earth comes to an end.**

Google: < red giant stars >

According to some estimates, **all life on Earth will end after** approximately **220 to 250 million years** – when the Sun becomes a red giant star swallowing or coming very close to everything in Earth's orbit. Before that time all living things on Earth will be well cooked/burned/charred. **At about that same time this solar system has orbited once around the Milky Way center at a speed of 220 kilometers (135 miles) per second. As has been mentioned, that time period is called a galactic year, or a cosmic year.** *Therefore, the life expectancy of this planet Earth is less than one galactic year.*

Google: < period of sun's orbit around the galactic center >
            < the sun – the future of us >
            < the milky way galaxy >

28.4. IF NOT GOD, WHO, OR WHAT DID IT ALL?

Somehow 'our' life-giving Sun, this Earth, life on this Earth, this whole Solar system, this Milky Way Galaxy and the whole universe came into existence.

Nobody but God/Creator knows all the details and never will. Even if the details could be known, there is not enough time for humans to pay attention to all those billions of details. Remember that one-year has only 31.5 million seconds. There will always be something new to learn for us and for all the generations to come.

The following example shows how limited we are. Remember that in the Milky Way Galaxy alone, there are over 100 billion stars, and that there are 3 billion seconds in one hundred years! How much about any new thing can one learn in one second, 24 hours a day, 365.242 days a year? If a centenarian sleeps eight hours a day, she/he has only 2 billion waking seconds. A person cannot learn very much in one second.

During the last few years and decades, the human knowledge and understanding of the universe, this solar system, this Earth and life on this Earth has advanced at greater speed than ever before. But the fact remains that no human being can ever know all the details of creation/evolution/life even on this planet Earth. Limitless studying and research remains to be done.

Every month, numerous researchers, scientists, engineers, medical doctors and professionals in many areas of science, technology, agriculture, manufacturing, research, development, industry, arts, etc. are increasing human knowledge and understanding of the universe. The list includes 'our' Sun, 'our' Solar System, planet Earth, life and living on this Earth, our health, wellness, illness, millions of applications in sciences, agriculture, engineering, technology, manufacturing, commerce, travel and daily human comforts and needs. **As the result of these developments, in many respects most of us have better, more comfortable, healthier and more abundant lives than the kings had one century ago.**

For more complete lists of the diversity of our lives, think of consulting the Yellow Pages of a large city telephone directory or Graduate School Bulletins of universities/schools/colleges. They will bring a good description of our many available opportunities/possibilities here on Earth.

Published scientific literature is continuously expanding. Abstracts alone in many fields, such as Physics, Chemistry, Medicine, Earth Sciences to mention only a few of many areas, annually cover thousands of new pages. There is 'so much' to know/learn, and 'so little' time available!

This book will be helpful providing general information, which can be easily understood about items concerning our home planet Earth with the help of the many, often updated Internet sites.

Remember that something made all this diversity possible. All honors for all of it in the final count go to the Creator/Whomever, not humans. Most of us take numerous 'things' for granted on this wonderfully balanced, absolutely unique planet Earth, the unique place for all human life in the entire **known** universe.

28.5. UNANSWERABLE ETERNAL QUESTIONS

Questions about our Solar System, such as how, what, when, where, why, will remain eternally without complete answers for us. We can/will/want to learn more every day/week, but the simple fact is that complete knowledge on everything cannot be reached by humankind.

There are many things in life we know for sure. Also, there are many things in life we do not know. We don't know all the details, nor who or what made/created life. We don't know everything about this planet Earth, although we do know quite a bit. **The Creator was much greater and more capable and ingenious than what we can even imagine**. Here one runs into questions of beliefs, religion and faith, where many opinions exist and differ. **No particular faith/religion, or person can have everything figured out universally correctly, although there are persons and groups of people who think that they have it all figured out correctly.** Scientists are more willing to admit that they don't know everything than some people of various religions.

Humans will never be able fully to understand/explain everything in the universe, not even everything on this Earth nor in this Solar System. However, **we are learning more every week, or at least, we should be. The areas of available learning, research and understanding are limitless for all of us.** We ought to broaden our horizons by pushing the frontiers of ignorance further away from us.

29. Some Details of Earth's Motions

Together with this Earth, **we are moving constantly in many ways.** This solar system orbits the center of the Milky Way Galaxy at a speed over 220 kilometers (135 miles) per second.

Google: < rotation of milky way galaxy >

Earth is orbiting all the time at a speed of 30 kilometers (18.6 miles) per second in its almost circular orbit around the Sun.

Google: < earth's orbit >

Earth spins once a day around its North – South axis running between the South and North poles. Maximum rotational linear speed of Earth's surface is 0.463 kilometers (0.288 miles) per second at the Equator. At the 60-degree latitude this linear speed and also the centrifugal acceleration is one half of the equatorial values. This rotational spinning causes the centrifugal acceleration, which at the equator reduces Earth's gravitational pull by 0.34 %, less at higher latitudes and by zero at the poles. **The (vector) sum of the pull of gravitation and the centrifugal acceleration is called gravity. The direction of the gravitational pull is nearly toward Earth's center and the direction of the centrifugal acceleration is in the plane of the parallel circle of latitude, perpendicular to the North – South spin axis.**

Google: < earth's rotation >

Earth rotates around its North-South spin axis giving us days and nights. Earth with its tilted (23.5 degrees) equatorial plane to Earth's orbital plane = Ecliptic plane = Ecliptica, produces the annual seasons: Summer, Autumn, Winter and Spring. **There are no other reasons for the existence of days and nights and the annual seasons.**

Google: < earth's four seasons >
             < astronomy dictionary >

Recall that the **average radius of Earth is 6371 kilometers = 3959 miles.** Earth is **almost spherical in shape** so that its polar radius is only 0.3 % shorter than its equatorial radius. The words used for Earth's general approximate size and mathematical shape are: **Rotation Ellipsoid, Ellipsoid of Revolution, or Oblate Spheroid.** A well-chosen Earth Ellipsoid has the same volume as the average (mean) sea level surface (=Geoid) with its continuation under the continental grounds. Topographical elevations are measured up from the Geoid and oceanic depths are also measured down from the Geoid.

The surface of the **mathematical** Rotation Ellipsoid is obtained by rotating any meridian ellipse once around its North-South axis.

Recall from earlier that the deviation of the shape of the meridian ellipse from a perfect circle is expressed by a number, which is called the Flattening of Earth. Earth's flattening is f = 0.0033528. Flattening of a perfect sphere is exactly zero. Flattening is the ratio obtained by dividing the difference between equatorial radius and polar radius by the equatorial radius. The equatorial radius is approximately 6378 kilometers = 3963 miles long, and the polar radius is 6357 kilometers = 3950 miles long. The circumference of Earth is approximately 40,000 kilometers = 24,855 miles. The lengths of the circumferences of the meridians are slightly shorter than the length of the Equator. All of them are approximately 40,000 kilometers = 24,855 miles long. Millimeter and few meter accuracies are not necessary for this book. The linear spinning (rotation) rate is at its maximum on the Earth's equator = 463 meters per second = 1520 feet per second = Mach 1.4. The linear spinning (rotation) rate is at its minimum at the North and South Poles, where it is zero.

Google: < wgs-84 >
             < geodetic reference system 1980 >
             < mach number >
http://www.grc.nasa.gov/WWW/K-12/airplane/mach.html

29.1. DAILY ROTATION OF EARTH

Earth rotates approximately 361 (360.986…) degrees around its North-South spin axis in 24 **mean solar hours** or in 24 hours of standard time with respect to 'our' Sun. **Standard or civil times** are the ones good civil clocks keep. Earth's rotation produces **days and nights. Universal Coordinated Time (UTC) is for zero-degree longitude** of Greenwich, England corrected for leap seconds. One hour time zones (actually time sectors) are 15 degrees wide in longitude, which fact originates from 360 degrees = 24 hours. For instance, five time zones, or 5 x 15 = 75 degrees west of Greenwich, England, is the Eastern Standard Time (EST) zone of USA for 75 degree West longitude.

The duration of time it takes Earth to rotate exactly 360 degrees is called one sidereal day. People working in areas of astronomy and Earth satellites use sidereal time more than the general public. A sidereal day is approximately 4 minutes shorter than the 24 mean solar hour day. Those clocks, which are 'gaining' approximately 4 minutes per day, run almost like the sidereal clocks.

Earth's spin axis is not completely stationary within the body of Earth. Those minute wobbles of the Earth's spin axis will not be discussed in this book.

Google: < **usno master clock time** >
             < **sidereal time** >
             < **wobbles of earth's spin axis** >
             < **leap seconds** >

29.2. EARTH'S ANNUAL REVOLUTION (JOURNEY) AROUND THE SUN

The velocity of this Earth 'flying' through the emptiness of the space surrounding 'our' Sun in its annual orbit is on the average 30 kilometers = 18.6 miles per second. If an object near Earth's surface would be flying that fast at sea level, its speed would be 88 times the speed of sound or 88 Mach. Sounds do not travel in vacuum, and therefore, Mach-numbers do not really exist in space. However, comparisons, such as above, can be made.

Earth's orbital velocity is approximately 30 times as fast as the muzzle velocity of a bullet coming out of a military rifle. If a rifle bullet would be going as fast (with respect to the ground) as Earth is 'flying' (orbiting), the kinetic energy (killing power) of the bullet would be 900 times (30 x 30 = 900) as great as that of a normal bullet coming from a muzzle of a military rifle.

Google: **< kinetic energy >**
              < speed of sound in air >

Due to the slightly elliptical shape of Earth's orbit around the Sun, the orbital speed is slightly faster in January than it is in July according to Kepler's Laws (Second Law) of the planetary motions. Earth is nearest to the Sun (at perihelion) in January and farthest (at aphelion) in July. One round trip of Earth around the Sun lasts one year.

The inner planets (nearer to the Sun than Earth) Venus and Mercury orbit at greater linear speeds than Earth, and their round-trips around the Sun last less than Earth's 365.24...days. **The outer planets Mars, Jupiter, Saturn, Uranus, Neptune and Pluto orbit at slower linear speeds than Earth,** and their trips around 'our' Sun last longer than Earth's 365.24.... days.

Google: **< kepler's laws >**
              < johannes kepler >

Johannes Kepler (1571-1630), a great scientist, was the first person to discover the basic Kepler's Three Laws governing planetary and satellite motions around central bodies (the Sun, planets). **Kepler's Laws have been valid and in full operation for eons before the appearance of humans on this Earth.** Therefore, even **Kepler's Laws are really God's/Creator's Laws, laws of nature, or laws of Physics, as they are often called.** Of course, honors of the discovery/invention should go to Johannes Kepler for being the first person clever enough to discover and publish those three laws of planetary motions, which have been in full operation for millions of years.

29.3. RECAPPING EARTH'S MOTIONS WE SHARE EVERY DAY

1. As was mentioned, there are several motions/velocities this Earth has in the known universe.

2. **Even the whole Milky Way Galaxy is moving** with respect to many nearby galaxies at a speed of approximately 300 kilometers (200 miles) per second.

Google: **< speed of the milky way in space >**
              < speed of the sun >
              < speed of the earth in orbit >

3. With 'our' whole solar system we are **moving (orbiting) around the center of the Milky Way Galaxy** at a velocity of approximately 220 kilometers (135 miles) per second, as stated previuosly. The duration of the total 360-degree orbital period around the Milky Way center is called the Sun's (and Earth's) **Galactic Year**. Stars nearer to the Milky Way center have greater linear velocities and shorter Galactic Years than stars at greater distances from the galactic center. **This is in resemblance to the Kepler's Second Law as applied to this solar system, where the inner planets (from Earth) similarly have greater orbital speeds and shorter orbital periods (their years) than the outer planets.**

4. Earth is moving in its **annual orbit around the Sun at** approximately **30 kilometers (20 miles) per second.**

5. 'Our' Moon orbits this Earth at an average speed of approximately one kilometer (0.6 miles) per second. The Moon's gravitation pulls Earth by small amounts in (new moons) and out (full moons) of its Keplerian solar orbit ellipse perturbing (disturbing) Earth's solar orbital ellipse.

6. Earth's daily spinning around its N-S axis in the **gravitational field of the Moon and Sun causes Ocean and Earth Tides heaving the ocean waters, whole continents and ocean floors up and down every day and night over a range of up to 40 centimeters (16 inches).**

Earth's daily spinning around its N-S axis in the **gravitational field of the Moon and Sun** *also causes* **the better known Ocean Tides sloshing the ocean waters around in the ocean basins.**

Earth's daily spinning around its N-S axis in the **gravitational fields of 'our' Moon and Sun** *also creates* **small** atmospheric tides, which can be observed as tiny systematic air pressure changes on top of atmospheric pressures of 'normal' weather systems.

At many places on the Earth, we are getting these nice up to 40-centimeter (16 inch) systematic up and down heavings of the ground and small air pressure changes which our bodies cannot sense!

To get an idea of some commonly encountered velocities of earthly travels, a few examples follow.

The Earth travels 30 kilometers (18.6 miles) per second in its orbit around 'our' Sun. Consider traveling across the USA from coast to coast at 30 kilometers (18.6 miles) per second. Covering this distance, which is approximately **3100 miles (5000 kilometers), would be traversed in less than three minutes.**

Google: < the datastore – planets >

Our Moon, which is also an Earth satellite, travels approximately one kilometer (0.62 miles) per second in its monthly orbit around this Earth. Together with the Earth, the Moon travels once a year around 'our' Sun.

Near the time of full moon, when the Moon is generally on the far side of Earth from the Sun, the Moon travels in the same general direction as Earth, and its speed is then approximately 31 kilometers (19.3 miles) per second with respect to the Sun. At the time of the near new moon, when the Moon is generally on the same side from Earth as the Sun, the Moon travels in the general opposite direction of Earth's orbital motion. Its speed is then approximately 29 kilometers = 18 miles per second with respect to the Sun.

The Moon's orbital plane (around Earth) is tilted (inclined) by approximately a five-degree angle to Earth's orbital plane around 'our' Sun. This means that during any one day the Moon can cross the meridian no higher or no lower than approximately five degrees from the spot where the Sun was at the previous high noon. If 'our' Moon would orbit in the ecliptic plane, the Moon's inclination would be zero (instead of approximately 5 degrees). Then there would be a solar eclipse and also a lunar eclipse every 'moon' (month).

Google: **< tilt of the moon's orbit >**

As has been mentioned, the **Sun with all of its planets/comets/asteroids travels approximately at a speed of 220 kilometers (135 miles) per second** around the center of the Milky Way Galaxy. It takes 'our' solar system over 200 million years for that round trip. **In addition, the whole Milky Way Galaxy is also traveling at a similar velocity with respect to other distant galaxies.** The whole Milky Way Galaxy may have other movements to its 'rolling along' such as wobbling of its central main disk.

The nearest spiral galaxy similar to 'our' Milky Way Galaxy is the Andromeda Galaxy at a distance of 2.5 – 3 million light-years from us. The distance between the Milky Way and Andromeda Galaxies is decreasing by approximately 300 kilometers (200 miles) per second. **This means that if *our distance* at the present time to Andromeda is exactly 3 million light-years, then after an elapsed time of one thousand years, the three million light year distance has 'shrunk' to 2,999,999 light years.**

If the mentioned distance keeps decreasing, the two galaxies may collide after a few billion years at an increased speed due to the increasing mutual gravitational pull as the distance gets shorter. The estimate (2007) for this galactic merging is 6 billion years in the future. It is likely that only a few stars will collide. Recall that the spacing between stars is huge (several light-years in 'our neighborhood of the Milky Way') and that our Sun's diameter is only 3.6 light-seconds. However, there will be many interactions in gravitation and radiation between many stars in the merging galaxies.

Google: **< andromeda galaxy >**
                  < interacting and merging galaxies >

29.4. EXAMPLES OF SOME SLOWER EARTHLY SPEEDS

The following is an **example of a somewhat slower speed**, which still is pretty fast for most of us. If the **winning sprinter runs one hundred meter dash in ten seconds**, his average speed is ten meters per second or one centimeter in one millisecond (0.001 second). If the second sprinter is ten centimeters (four inches) behind the winner, his time for the 100-meter dash will be 10.010 seconds. Further, because one hour has 60 x 60 = 3,600 seconds, it follows that, if the winning sprinter could continue running at his winning speed for one hour, he would travel 3,600 x 10 meters = 36 kilometers (22 miles). The winner's average speed in his one-hundred-meter dash was 36 kilometers = 22 miles per hour.

Imagine what a jolt a sprinter would get if he ran at 22 MPH into a concrete wall! If the deceleration is uniform from 22 MPH to a stop in two feet (60 centimeters), the deceleration is approximately 8.5 G for 0.12 seconds. Because the runner's arms are not strong enough to push 8 times his weight for braking his speed enough, the final jolt will be much more than the 8.5 G. **Please have respect for a dangerous jolt you will receive by driving a car at just 22 MPH into a tree!**

Typical velocities of ground motions (continental wanderings, land uplifts, subsidences and earthquakes) are of the order of from one millimeter to a few centimeters per year. Earthquake movements can be much greater. Among these large scale slow motions are continental wanderings such as the general widening of the Atlantic Ocean and the general narrowing of the Pacific Ocean.

Google: < **measuring land subsidence from space** >

There are areas on Earth where the ground and the sea floors systematically rise or sink by a few millimeters per year. For instance, **large areas of Canada and Scandinavia are still rising as the result of the melting of up to 3-kilometer (2-mile) thick continental ice sheet after the last ice age some ten thousand years ago**. There are also some other reasons why the ground moves vertically. One such reason is pumping out lots of water from the ground.

According to some estimates, ice sheets over Scandinavia (and Canada) pressed the ground down by approximately **up to 900 meters (2950 feet) during the last ice age.**

The 15 to 80 kilometers (9 to 50 miles) thick **Earth's Crust is floating above the denser underlying more or less viscous Mantle, like ice floats on water**. The **floating equilibrium of the Crust over the Mantle is called isostatic equilibrium**, where a thicker Crust, such as in mountainous areas, has sunken deeper into the Mantle than a thinner Crust under low lands or under the ocean floors. According to Archimedes' principle, this isostatic floating of Earth's Crust over the denser Mantle underneath can be compared to floating icebergs in the oceans.

The land uplifts and land subsidences (sinkings) are called isostatic equilibrium adjustments, which are ongoing at many locations usually at speeds of a few millimeters per year.

As the ice-age thickening **ice sheets pressed the ground (Crust) deeper into the underlying upper Mantle**, the disturbed Mantle material generally flowed slowly 'out–of–the–way' toward surrounding areas. After the ice sheets melted, the extra loads on the affected ground areas were removed, and the ground is still rising every day, now some 8,000 years after most of the ice melted.

The affected Mantle material is now slowly returning toward its original locations causing the Crust (ground) to rise toward its original pre-ice-age topographic elevation and equilibrium. For instance, due to this uplift, the land area of Finland is increasing by approximately twenty acres (ten hectares) per year.

At first, the rising of the ground was much faster than it is now. The present uplift rates in Scandinavia are up to 9 millimeters (0.35 inches) per year. At some places in Western Canada the land is rising by a few centimeters per year.

Google: < earth's interior >
            < isostasy >
            < last ice age >
            < archimedes' principle >
            < isostatic equilibrium >
            < isostasy and gravity >
            < ice ages >
            < milankovitch cycles >

In the past Earth has experienced several ice ages, and **it is likely that there will be new ice ages in the future.**

Estimates for the appearance of the next full-blown ice age range from 2,000 to 100,000 years. According to some scientists, the next ice age has already begun.

When that time comes, it will be likely that most of Alaska, Canada, Scandinavia and Northern Russia will be covered again by 1 –3 kilometer (1 –2 mile) thick ice layer (ice sheet). Sorry about that, you good people who live there and your unborn future generations! This book is only a messenger!

30. ORBITS OF SOLAR PLANETS

Recall that Earth's orbital plane around the Sun is called the **plane of the ecliptic or Ecliptica**. The imaginary straight line, one AU long (radius vector) from the center of the Sun to the center of the Earth-Moon system (barycenter), sweeps (creates) this plane surface in one year. It is a very stable plane. Year after year the Earth (and the radius vector to the Sun) repeats the same orbit with only very minute variations, which are not described in this book.

It is also interesting to note, that all **other eight solar planets** – with an exception of the smallest planet Pluto with its 17-degree orbit inclination to Earth's orbital plane **and even 'our own' Moon – have orbital planes, which are tilted (inclined) only by a few degrees with Earth's orbital plane (Ecliptica).**

Therefore, all major objects of this Solar System form a relatively tight disk-shaped entity. It is a very stable plane in this solar system. Some comets and extra solar objects may move at higher inclinations crossing the plane of the ecliptic with inclined angles all the way up to 90 degrees.

Google: < orbits of solar planets >
                 < orbital properties of the solar system >
                 < the nine planets >

'Our' Moon's average distance from the center of the Earth is 1.2 light-seconds = 60 times Earth's radius = 0.002 Astronomical units.

The average distances of the nine solar planets from the Sun are as follows.

Mercury: 0.387 AU
Venus: 0.723 AU
Earth: 1.000 AU, or 500 light-seconds

Mars: 1.524 AU

Jupiter: 5.203 AU

Saturn: 9.6 AU

Uranus: 19.2 AU

Neptune: 30.1 AU

Pluto: 39.5 AU = 19,750 light-seconds, or it takes light 5 hours 29 minutes 10 seconds to travel between Pluto and the Sun.

From these numbers one can see that the **minimum possible distance to Mars from Earth is approximately 0.5 AU** (250 light-seconds) when Earth and Mars are on the same side of the Sun. The **maximum possible distance is approximately 2.5 AU** (1250 light-seconds), when Earth and Mars are on the opposite sides of the Sun.

For more accurate considerations, one would need to take into account that the tilt (= inclination) of the orbital plane of Mars with respect to Earth's orbital plane (Ecliptica) is 1.85 degrees. An approximation of this 1.85-degree angle is formed between a 12-inch ruler and the tabletop by lifting one end of the ruler 0.4 inches (one centimeter) above the tabletop.

Google: **< jpl solar system dynamics >**

One can see the inclinations of the solar planets by clicking on the JPL-site on '**Mean Orbital Elements**' under Planets. Pluto's inclination is 17 degrees, Mercury's inclination is 7 degrees and the inclinations of all other planets are less than 4 degrees.

Clicking on '**Orbit Plots**', under Planets, and there for 'Inner Solar System Orbit Diagrams' and then for 'Outer Solar System Orbit Diagrams', good **orbit diagrams for inner** and **outer planets** appear on the monitor-screen. The general distribution of asteroids and comets in this solar system are also depicted.

31. Time Keeping, Basic Units Of Time

One good benefit for having time is that everything does not happen at once. We have seconds, minutes, hours and years, one at a time. From where do they originate? What is involved??

Time and time keeping are not the easiest things to fully understand because time **involves Earth's slightly elliptical orbit around the Sun in the plane of the ecliptic and Earth's rotation (spinning) around the N-S spin axis with respect to the Sun and also with respect to the distant ('fixed') stars**.

Angular directions from Earth to all 'fixed' stars (except to 'our' Sun) are practically constant throughout a year (and years). But because the Sun is so very close to us and because we orbit it once a year, we 'see' all the horoscope (Zodiac) constellations one at a time being behind the Sun.

Google: < **annual trigonometric parallax** >
             < **zodiac constellations** >

The original units of time were derived from Earth's spinning around its North – South axis and from Earth's annual orbiting 'our' Sun. **Daily 360-degree (diurnal) rotation (spinning) of Earth defines the length of one particular day, and the annual 360-degree orbiting of Earth around the Sun defines the length of one particular year**. Modern measurements are so accurate that small variations from the average (mean) values are known to a small fraction of one second.

To "Walk Through The History Of Time Keeping", go to:
http://physics.nist.gov/GenInt/Time/biblio.html

When there, click on the thumbnail images from **Ancient Calendars, Early Clocks, A Revolution in Timekeeping, The "Atomic" Age, World Time Scales and Time Zones, NIST Time Services** to **Bibliography**.

In the 'Ancient Calendars' section, one can read that the Moon was the time-keeper for many ancient 'civilizations'. When for instance, the Bible 'tells' that Metusaleh lived to be some 900 years old, the original intention **may have been** to tell that he lived to be 900 months (moons) old, which in solar years would be approximately 75. In those ancient times, an age of 75 was very unusual and respectable age for any person.

A good description of the 'Painful History Of Time' can be read at the following two sites:

http://naggum.no/lugm-time.html

http://www.time.gov/exhibits.html

Find official U.S, standard time to the nearest second 24/7/365 at:

http://nist.time.gov/timezone.cgi?Eastern/d/-5/java

Find the stability of modern timekeeping at NIST at:

http://tf.nist.gov/general/glossary.htm

As is well known, minutes have 60 seconds, and hours have 3,600 seconds and further – maybe not so well known – that **mean (average) solar years have 365.2419878 days or 31,556,907.75 seconds.**

Try to remember that one year has 31.5 million seconds. Human hearts beat about that many times per year! It has been written/said that when Galileo Galilei (1564-1642) was studying the swinging times of short and long pendulums (church chandeliers), he used his pulse for timing. There were no wristwatches some 400 years ago.

Google: **< definition of one second >**

               < nist time and frequency division >

According to the present definition, one second of time is the duration of 9,192,631,770 periods of the radiation corresponding to the transition between the two hyper fine levels of the ground state of the cesium 133 atom. **Most of us have to leave that for the experts.**

The modern highly accurate definitions for the meter and the second became possible after atomic clocks were developed and built. High accuracy timing is very important in many facets of modern science and living, although this high accuracy is not directly obvious in routine every-day activities. **Among**

the numerous modern applications of the highly accurate atomic clocks/ timing/time keeping are many instruments, computers, satellites and the GPS (Global Positioning System). Be glad and grateful that these items are being taken care of by the experts working in the field!

The average duration of (approximately) one 361-degree (not only 360) daily rotation of Earth around its North-South spin axis with respect to the Sun (we live by the Sun, not by the stars) throughout a year determines the length of a mean (average) solar day, which has 24 mean *solar* hours. Each hour has 60 minutes or 3600 seconds. The universal coordinated (mean) solar time is the time our good clocks keep with slight corrections (leap seconds) determined by atomic clocks timing the very slight variations in Earth's rotation rate.

The length-of-day changes (leap seconds) are caused by small sporadic/systematic mass movements in Earth's atmosphere, oceans, lakes, river waters and in the interior of the Earth. These mass movements change the Earth's angular momentum causing these small speed-ups/slow-downs in the Earth's spin rate. This can be compared to figure skaters being able to vary their spin rates by moving their arms.

Google:　　< angular momentum >

Atomic clocks keep much more uniform time than Earth with its fairly constant rotation rate. The resulting very uniform time controlled by the atomic clocks is called Universal Coordinated Time or UTC. The order of letters in UTC comes from the French language. Atomic clocks reveal the small irregularities in Earth's rotation rate.

The time kept by Earth's daily rotation is called UT1. A few leap seconds have been added since 1972 as needed to the mean solar time UT1. No leap seconds have been subtracted. This means that the spin (= rotation) rate of Earth has slowed down by the amount described by those few leap seconds since 1972.

At present time the difference between UT1 and UTC is changing (Earth's spin rate is slowing down) at a rate of approximately 2 to 3 milliseconds per day, which makes an addition of one leap second slightly more often than once a year.

Google:　　< leap seconds >
　　　　　　< http:/aa.usno.navy.mil/publications/docs/AsA_history.html >

31.1. SIDEREAL TIME

One sidereal year has 366.242 sidereal days, whereas, the mean solar year has exactly one day less or 365.242 days. Prorating this per one day, one finds that in one 24-hour mean solar day (24 hours standard time), there are 366.242 / 365.242 times 24 = 24.066 sidereal hours = 24 hours 3 minutes 56.6 seconds, i.e. **sidereal time gains almost 4 minutes every day when compared to standard time.**

Therefore in *one mean solar day* (= 24 mean solar hours) as Earth orbits 'our' Sun, **Earth has to rotate 361 (360.98...) degrees** around its N-S spin axis, whereas **in *one mean sidereal day* (= 24 sidereal hours) the Earth rotates exactly 360 degrees** around its N-S spin axis.

The duration of exactly one 360-degree daily rotation of the Earth around its North-South spin axis determines the length of that particular *sidereal* day, called apparent sidereal day. To repeat, sidereal days are 3 minutes 56 seconds shorter than the mean solar days kept by our civil clocks, just because Earth has to rotate approximately 361 degrees with respect to 'our' Sun as it travels in its orbit approximately one degree per day. We live by the Sun, not by the 'fixed' stars. Sidereal time has one (of many) application with the communication satellites in equatorial 24-hour orbits. **There are no sidereal calendars, months or weeks. Each calendar day has a sidereal time for its every instant, which can be computed as needed.**

Communication satellites are used for many purposes (TV is one of them). Communication satellites are in *24-sidereal hour* equatorial orbits, not in 24-solar-hour orbits. If they would be in 24-hour mean solar time orbits, they would not stay 'parked' at one desired longitude over the Equator, but they would drift once around the globe in one year making them useless, or at least much more complicated to use.

Google: < sidereal day >

From where does the rotation of 361 (360.98564791) degrees per day come?

It comes from the fact that in one solar year, **Earth travels exactly 360 degrees from one point in its orbit to the same point one year later, but the**

number of solar days (Earth's rotations with respect to the Sun) **during that interval of time (in that one year) is 365.2419878. During the same year, there are 366.2419878 sidereal days,** or exactly one day more (one more 360-degree daily spin) when compared to one solar year.

Why must there be all these numbers?

God/Creator made it so, and there is nothing humans can do to alter or round off those numbers. As the accuracies improve in the coming years, only some last decimal numbers may be slightly updated. **This should make us aware what the experts in this timing area are doing.** Most of us don't need to dwell into these details, but it is useful to be 'in the know' of the real world where we live.

As was stated, in one sidereal day Earth rotates exactly one complete 360-degree rotation with respect to 'fixed' stars or distant galaxies. We live our daily lives by 'our' Sun, such as from high noon to the next high noon or from midnight to the next midnight. **We don't keep our civil time by some Milky Way 'fixed' stars or by some distant galaxies.**

The simple reason for the possible confusion in understanding the time and time keeping is the fact that in addition to its daily spinning, Earth also travels approximately one degree per day in its annual orbit around the Sun. **God/Creator made it to be so, and that is the way it is. We are very lucky to live on this Earth.** The decimal part 0.2419878 in the 365.2419878 is the reason for the existence of the leap years.

Google: **< earth's orbit around the sun >**
                 < leap years >

Another way to clarify the situation can go as follows. Keep in mind that in one day (24 x 60 x 60 = 86,400 seconds) Earth travels in its annual orbit approximately one degree-arc of its complete 360-degree annual orbit, or about 2.6 million kilometers (1.6 million miles) with its orbital speed of 30 kilometers (18.6 miles) per second. As a result, to get from one high noon (when the Sun is in the south crossing the meridian) to the next high noon, Earth must rotate approximately another degree more in addition to the 360 degrees for that meridian to transit the Sun again. This extra, approximately one-degree rotation takes about 3 minutes 56 seconds of mean solar time.

Because this time-concept is not very easy to grasp, here is still another look at the situation.

In one year, the approximately four-minute difference per day amounts (accumulates) to exactly one day. One solar year has **365.2419878 mean solar days**, meaning that Earth rotates 365.2419878 times around its North-South spin axis **with respect to the Mean (average) Sun**. Also in exactly the same time interval (of one complete orbit) **Earth actually rotates 366.2419878** times around its North-South spin axis **with respect to the 'fixed' stars**. Note once again that **the difference is exactly one day, or one sidereal year has exactly one more day than one solar year.**

Still another way to see this is the following. **Imagine that this Earth would be orbiting the Sun in its present orbit as it is, but it would not spin at all around its North–South axis.** Then for one half of the annual orbit, a meridian (your meridian) would be on the sunny side, and the other half of the orbit, your meridian would on the night side. In that case, the full year would have only one day and one night. **That one-day constitutes the mentioned one-day difference in the number of solar and sidereal days in one year.**

32. Compass Directions, Azimuths and Bearings

A zimuth is the angle in the horizontal plane that starts from North zero in a clockwise manner to 90 degrees East, to 180 degrees South, to 270 degrees West and again to 360 degrees North. Directions do matter when traveling somewhere. For example, one may want to know the compass direction (azimuth, bearing) in which a fence line, an airport runway, a highway runs or the azimuth (compass direction) from 'here' to Mecca in Saudi Arabia. Azimuth angle can be computed from 'here' (this station) to another point (the other station) from the coordinates of these two points (stations), if magnetic compass directions are not accurate enough.

Sometimes one needs to know where the compass directions are, whether one is walking, driving, sailing or flying. **Azimuths, bearings and compass directions** (north east, south, west and so on) **emanate from some one initial starting point on the horizontal plane. Azimuths are lines of direction** (usually somewhere between 0 to 360 degrees **from North clockwise) on the horizontal plane at the starting point.** Azimuths are usually expressed in arc degrees, minutes, seconds or in fractions of a second.

Google: < azimuths >
            < compass directions >
            < compass bearings on maps >

Azimuths – as do days and nights – originate from the way Earth rotates (spins). **Earth spins in the easterly direction around its North-South spin axis, and that basically sets the origin for East, North and all other directions from one point to another point.** Due to the daily spinning of the Earth, stars, Sun, planets and Moon rise, or go upwards in the eastern skies, and they set, or go downwards in the western skies without exception. Even the circumpolar stars, which stay above the horizon all the time, go up on the east side of the meridian, and they go down on the west side of the meridian.

A meridian ellipse runs in the North to South direction from the North Pole in the Arctic Ocean to any chosen point on Earth's surface. A celestial meridian is the extension of an earthly meridian plane to the sky all the way to the celestial sphere. The celestial sphere is considered to have an infinitely large radius. In earthly positioning by stars, only the directions to the celestial sphere matter, not the actual astronomical distances. However, in GPS (Global Position System), the distances to the GPS satellites are of crucial importance.

Azimuth is the (compass) direction in the horizontal plane starting from North clockwise. On most maps the direction to the north is straight up the map sheet. Azimuth to the North is zero degrees, azimuth to the East is 90 degrees, azimuth to the South is 180 degrees, azimuth to the West is 270 degrees and azimuth 359 degrees is one degree shy from the North on the west side of the meridian.

Airport runways have numbers, which give the azimuth of the runway to the nearest ten degrees. For instance, if a runway runs in an east-west direction, and an airplane is landing from the east, it touches down after flying over a large painted number 27. If an airplane is landing on the same runway from the west, it touches down after flying over a large painted number 09.

Azimuths (compass directions) by some automobile compasses are given to the nearest 45 degrees, such as N, NE, E, SE, S, SW, W and NW. In some precision surveys azimuths are needed to a fraction of a second of arc.

In some old applications in surveying and **on some old maps, azimuths may start from the south running clockwise from zero to 360 degrees.**

Sometimes azimuths or directions are (confusingly) described as bearings from North or South. For instance, on some maps azimuth 52 degrees can be marked as N52E (from North 52 degrees Eastward), or azimuth 203 degrees is marked as S23W (from South 23 degrees Westward). Trigonometric calculations are easiest done with calculators/computers by using azimuths starting from North, running from zero to 360 degrees.

Where is North? Azimuths are and were measured most commonly (in USA and Europe already 200 years ago and earlier) by observing the horizontal angle between Polaris, some other star, or Sun and some desired point on the ground. The azimuth of Polaris, another star or Sun can be easily computed for any instant of night/day for any point on Earth. Combining the computed azimuth of Polaris (or Sun) with the measured horizontal angle between Polaris (or Sun) and the 'desired' point on the ground, the azimuth of the 'desired' point from the measuring station can be obtained by a simple addition/subtraction of the two horizontal angles. Anywhere you go, you need the direction in which way to go!

An instrument used to measure horizontal/vertical angles is called a theodolite. Some older surveying instruments are called transits. A transit is not a good name for it! The telescope of some theodolites can be pointed to any azimuth and to any altitude from 0 to 90 degrees up and maybe some 45 degrees below the horizon.

Google: **< ngs faqs >**
            < product overview theodolite from wild heerbrugg >
            < geodesy products from wild heerbrugg >

Magnetic compasses can give only approximate general azimuths (map directions, bearings). They are useful for many purposes, but they are not accurate enough for precision work, such as determining property lines.

If, for instance, a GPS instrument measures the coordinates of two points, the azimuths between these points can be computed by the inverse problem.

Google: **< magnetic compass >**
            < gps positioning >

33. Geodetic (= Geographic) Coordinates

Positioning basically answers the question: **"WHERE ON EARTH IS IT"?**

Modern **Global Positioning System (GPS)** methods are in general use for **geographic positioning of points (getting geodetic coordinates: latitudes, longitudes, elevations, azimuths and distances between desired points)** on Earth's surface and in the air (airplanes, etc.) and also above the atmosphere (satellites). GPS methods are now usually the fastest and most economical positioning methods.

33.1. Latitudes on Earth

Geographic (geodetic) latitude of a point on Earth's surface is an angular distance to the parallel circle through that point along the meridian from the Equator north or south. We all know that latitudes range from zero to 90 degrees north or south. **Parallels of latitude are small circles of constant latitude.** They are circles with their centers inside the Earth on the North – South spin axis. Recall that the length of a meridian arc from Equator to either pole is 10,000 kilometers (6214 miles). One degree of latitude difference along the meridian is 111.1 kilometers = 69 miles = 60 nautical miles.

33.2. Longitudes on Earth

Longitude of a point on Earth's surface is the angular distances between two meridian planes, when one of them is the Prime Meridian plane going through the Greenwich Observatory in England. The other plane is the meridian plane through the point in question. The east and west longitudes are measured from Greenwich 0 to 180 degrees east or west. Sometimes the longitude is given/measured in an easterly direction 0 to 360 degrees from Greenwich making trigonometric computations easier. For instance, a longitude of 273 E is the same meridian as the 87 W meridian.

Recall that GPS instruments provide the fastest and most economical – but not always the most accurate – determinations of positioning for the latitudes, longitudes and topographical elevations – thanks to US Department of Defense and some smart scientists/engineers and the US taxpayers.

Astronomical latitudes and longitudes are the original results of classical positioning methods. There are many astronomical methods available to determine them. **Astronomical coordinates of a point may differ by a mile or so from the geographical (geodetic) coordinates**. The reason for this is the **irregular mass distribution inside Earth causing plumb line deflections in any azimuth**, usually by less than by one minute of arc. The plumb line deflections affect the leveling of astronomical instruments. The plumb lines are perpendicular to the mean (average) sea level = Geoid, which is a 'bumpy' surface approximating a perfect, mathematical Earth-ellipsoid of revolution. The 'bumps' are called Geoid Undulations, which usually are not greater than 150 meters (450 feet) over or under the Geoid.

Google: **< deflections of the vertical >**
< geoid >
< geoid undulations >

33.3. TOPOGRAPHIC ELEVATIONS ABOVE THE MEAN SEA LEVEL

Topographical elevation is the length along the plumb line from the mean (average) sea level (=Geoid) to the point on the ground. There are only a few land areas on Earth where the ground surface is below sea level.

Before the GPS receivers became available, accurate determinations of ground elevations were performed using so called spirit or automatic leveling instruments, or sometimes trigonometric leveling. Those survey methods are slow, expensive and tedious, as well as the triangulation and trilateration methods for latitude and longitude determinations of pre-GPS era.

Before the GPS-era, theodolites and electronic distance measuring instruments (electronic ones after 1950) were used to measure the triangle angles/sides of the triangles in triangulation and trilateration nets. Those methods were also limited to ground stations only. The GPS has brought a big 'sigh of relief' to most latitude/longitude/elevation determinations, and also in many other areas where three-dimensional locations of points on Earth's surface are required/desired.

Google: < national geodetic survey >
             < highs and lows. topography and isostasy >
             < us naval observatory gps operations >
             < us navstar global positioning system >
             < north and south latitudes and longitudes >
             < topographic elevations >
             < understanding topographic maps >
             < cartographic maps >
             < trigonometric leveling >

33.4. CELESTIAL METHODS FOR POSITIONING

Still in the 1950's, many distances to islands and over the oceans were not well known. For instance, the width of the Atlantic Ocean and the locations of many islands were not known much better than to the nearest mile (kilometer). After World War II a few distances across the Atlantic Ocean were measured to within approximately 100 meters (300 feet) by using Solar Eclipses, which occur very seldom. Electronic distance measuring methods (Loran, Shoran, Shiran, US Navy Satellite Transit System) were developed after World War II. Astronomical point positioning was in general use before 1950. It was the most accurate method over long distances, i.e. for points beyond the visible horizon.

Google: < loran >
             < geodetic surveying 1940-1990 >

Possibly the first positioning method using star backgrounds for man-made flashes in the sky was when the targets by camera-type magnesium flares (flashes) were carried to desired altitudes by weather balloons. The individual magnesium flashes were simultaneously photographed from a number of stations against the star backgrounds, which star backgrounds are different from various stations. This method was invented, tested and used by Professor Yrjö Väisälä in Finland starting around 1939 – 1940.

He got his idea by watching from Turku, Finland the flashes of the Finnish anti-aircraft-artillery-shells exploding against the star-background over Helsinki during Finland's Winter War in 1939-1940, some 200 kilometers = 120 miles from Helsinki. Both cities are in southern Finland. See the map of Finland at:

http://virtual.finland.fi/netcomm/news/showarticle.asp?intNWSAID=27068

By measuring the positions of the flare(s) and the known background stars from the photographic plates taken at each station, the ground coordinates of the stations (from where the photographs were taken) could be computed over longer distances than targets on the ground provided in earlier years.

When the satellite methods became available after 1957, several improved optical Satellite Triangulation Methods were built, tested and used mainly by US government organizations.

The basic Väisälä's StarTriangulation method was used in the 1960's and later by NOAA (National Oceanic and Atmospheric Agency), NASA and DOD (Department of Defense) with some elaborate tracking cameras on ground stations around the globe. The light-flashes were from special orbiting satellites. The satellite Triangulation System produced ground station coordinates within a few meters (feet) even across the oceans and from island to island. The whole Earth's surface was covered by triangles as large as the conterminous United States.

Google: < satellite triangulation by noaa >
              < satellite triangulation >
              < us navy satellite transit system >

http://www.esg.montana.edu/gl/usa/25.00024.000065.00050.00012000676a3.
html Click anywhere on this map and get approximate coordinates, elevations and
some other information for that point, even the elevations for the Great Lakes,
Salt Lake in Utah, Pikes Peak in Colorado, Lake Okeechobee in Florida, etc.

33.5. GPS, THE GLOBAL POSITIONING SYSTEM

Precise classical celestial positioning methods are suitable only for stations on
'solid' ground'; less accurate sextants can be used on ships and airplanes. Due to
plumb line deflections (the Geoid = mean sea level is bumpy), the 'precision' of
such classical positioning methods is often off by a mile.

The US Navy Doppler Transit System was a global electronic positioning
method, which could be used also on ocean-going ships.

Google :   < sextants >

              < celestial navigation data >

              < us navy navsat satellite system >

The GPS (Global Positioning System) was the next positioning system developed
by/for US Department of Defense (DOD) by many organizations and industries.

The original name for GPS was **Geodetic Positioning System**. The GPS system can produce station coordinates within a few centimeters (inches), in some
cases even to millimeters. GPS receivers can be used on many types of moving and stationary platforms such as Earth satellites, airplanes, missiles, bombs,
ships, cars and other vehicles.

Google: < the global positioning system >

**After the development and building of accurate atomic clocks, sat-
ellites with onboard atomic clocks, necessary electronics, transmitters,
antennas, rockets to put them in suitable Earth-orbits and continuous
determinations of very accurate GPS satellite orbit positions, the GPS
system became possible.** There is at least one atomic clock aboard every
(about two dozens of them) working GPS satellite.

Recall that the personnel from the United States Department of Defense, together with some scientists at a few universities and scientists and engineers in some electronic industries, provided the brainpower. Funding has been by American taxpayers.

The first Block 1 Navstar Global Positioning System satellite was launched on Feb. 22, 1978 from Vandenberg AFB, California. The event received little notice in the press. Twenty-five years later, **GPS has become one of the most successful and versatile high-technology projects of all time.** GPS has turned out to be one of the most important US government investments in space, creating over $30 billion a year in civilian GPS–related devices and services. It is also very important and useful for ships at sea: for navigation, for civil aviation, for surveyors, for all mapping and for all military services in many ways.

The Global Positioning System is a constellation of satellites that beam navigational data to anyone in the world with the proper equipment to receive it. The GPS satellites are, in essence, extremely accurate atomic clocks in the sky.

The GPS signals are so accurate that time can be determined within a millionth of a second. Locations on Earth can be pinpointed within 10 meters = 33 feet and in some applications within millimeters.

The GPS system is operational on worldwide basis benefiting all countries. The annual operational upkeep of the GPS system is a major DOD operation benefiting a large number of civilian and military users all around this Earth on land, sea, air and space.

Google: < us naval observatory gps operations >
              < awc military space >

The GPS is one of the all-time great inventions/developments. GPS has made airplane flying/landings safer, especially in bad weather. GPS is helpful for US Military in many ways on land, sea and air. GPS guides search teams in the open ocean to the once-found spot where, for instance, the Titanic or an airliner rests on the ocean floor. GPS has guided many rescue teams to places where their help was desired. **GPS is a 'dream-come-true' for many navigators, surveyors, mapmakers and the US Department of**

Defense. GPS is the proverbial 'magic black box' for locating points with accurate coordinates almost anywhere (signals can be blocked by obstructions, such as buildings) **on Earth.**

GPS has revolutionized/improved navigation, land surveying, map making and positioning on land, sea, air and in the surrounding space. One well-publicized (July 2002) application of GPS surveying was the ability to quickly drill a hole in the correct place on the ground to the mineshaft at Quecreek Mine in Pennsylvania to rescue nine trapped miners. GPS surveyors quickly found the spot on the ground to drill the rescue holes. These drill holes were approximately 4400 feet (1345 meters) from the mine entrance.

Google: < quecreek mine rescue >

33.6. ASTRONOMICAL POSITIONING

Astronomical latitude of a station (= a place on Earth's surface) is equal to the altitude (angular distance up from horizon) of the Celestial Pole, which for northern latitudes is in the direction (extension) of Earth's spin axis to the sky hitting the celestial sphere not far from the North Star or Polaris. At Earth's North Pole, the Celestial Pole is directly overhead at the zenith. Due to Earth's rotation, all stars make daily (diurnal) circles around the Celestial Pole.

Google: < daily motions of stars >

For example, anywhere at the 45N parallel of latitude on Earth, Polaris is in the north at approximately 45-degree altitude above the horizon. At the North Pole, Polaris is almost straight up overhead within one degree of the North Celestial Pole. Polaris makes a small diurnal circle around the North Celestial Pole. At zero-degree latitude at the Equator, Polaris is half the time slightly above the horizon in the north. The other half of its diurnal circle, Polaris is below the horizon (ignoring the atmospheric refraction).

Google: < earth as an ellipsoid >
             < astronomical positioning >

Astronomical longitude is the angle between the local and Greenwich meridian planes. Astronomical coordinates refer to the bumpy Geoid (mean sea level). Geodetic (geographical) longitude refers to the smooth, mathematical rotation ellipsoid.

Astronomical latitude is the angle between the equatorial plane and the plumb line at the station. The plumb line is perpendicular to the bumpy Geoid. **Geodetic (geographical) latitude refers to the smooth mathematical rotation ellipsoid.** It is the angle between the equatorial plane and the ellipsoid normal at the station.

Note that the ellipsoid normal is perpendicular to the smooth mathematical Ellipsoid of Revolution, and that the plumb line (perpendicular to the Geoid = Mean sea level) may deviate from the Ellipsoid normal in any azimuth by up to 150 seconds of arc, which can result in approximately 4.5 **kilometer (2.8 mile) errors in latitudes and longitudes on the ground.**

Classical/historical astronomical latitude and longitude determinations were/are performed by theodolites, meridian transit instruments, astrolabes and zenith telescopes together with the most precise timing devices available. Accurate timing for the instants of the observations and precise star coordinates were/are obtained/computed for the instant of star observation from accurate time sources and star catalogs.

Google: **< usno 6-inch transit circles >**
               < theodolites >
               < meridian transit instruments >
               < western astrolabes >
               < the united states naval observatory >
               < the pzt zenith tube website >
               < astronomical star catalogs >

34. CELESTIAL COORDINATES, DECLINATION AND RIGHT ASCENSIONS

The ancient star constellations give only the general area in the sky for any object 'out there'. Constellations are not precise enough to get a desired object in the field of view/crosshairs of a telescope.

It is possible to get the Sun (using suitable filters), Moon and bright stars/planets in the field of view of many telescopes just by using the gun-sights on the telescope, but many desired objects are not bright enough to see by the naked eye through the telescope gun-sights. Therefore, getting the faint galaxies, stars, planets, moons and other objects 'out there' in the crosshairs, many detailed items are needed about the object's coordinates on the celestial sphere together with the time of day and the geographical location of the telescope. **Star coordinates on the celestial sphere are the required items to get the desired celestial object in the crosshairs of the observing telescope.**

Declinations and Right Ascensions are important celestial coordinates for astronomers, space scientists and other specialists to locate, observe, measure, and photograph desired celestial objects.

It takes expertise to do all that! Without having declinations and right ascensions one does not get very far in getting most of the celestial objects in the crosshairs of a telescope.

Many stores that carry globes also carry models of celestial spheres, which may be informative three-dimensional models for visualizing celestial coordinates. Also note that most Earth-globes have their spin axes tilted by 23 degrees, which is the obliquity of Ecliptica, i.e. the 23-degree angle between the planes of Earth's equator and Earth's orbital plane around the Sun.

187

Google: < diurnal circles >
               < coordinates on celestial sphere >
               < celestial coordinate systems >
               < celestial coordinates >
               < definition of right ascension on celestial sphere >
               < precession and nutation of earth's spin axis >
               < sidereal time >
               < what does obliquity mean? >

34.1. DECLINATIONS

Celestial sphere is used to specify directions to celestial objects in a similar manner as latitudes and longitudes are used to specify locations on the spherical/ellipsoidal Earth's surface. The latitudes on Earth are measured from Earth's Equator north or south in angular units (degrees, minutes, seconds and fractions of a second). One component of the directions to the galaxies, stars, Sun, planets, moons and other objects 'out there' is measured on the celestial sphere in angular units from the celestial Equator north (positive numbers) or south (negative numbers) from zero to 90 degrees. **These angular distances from the celestial Equator are called declinations.** Declinations of celestial objects on the celestial sphere are similar to latitudes on Earth.

Declinations (in degrees, minutes, seconds and fractions of a second) of stars/objects give the angular distance north or south from the celestial Equator. **The Celestial Equator** is the extension of Earth's equatorial plane to the celestial sphere. It is, therefore, directly above points on Earth's Equator, and its declination is zero. Stars north of the Equator have positive declinations, and the southern stars have negative declinations. Declination of Polaris is listed daily in many star catalogs for zero hour UTC (= beginning of the day at Greenwich, England). The declination of Polaris changes slowly and it is now approximately +89.25 degrees.

Earth's daily rotation causes all stars to describe their own diurnal circles on the celestial sphere. These diurnal circles have the declination of the star that described it. Every diurnal circle has its own almost constant declination. 'Almost constant', because the minute variations in declinations must be computed for the instant of observation/measurement when doing precise astronomical work.

Polaris, the North Star with its declination of slightly over +89 degrees, is within one degree to the North Celestial Pole for the rest of our lifetimes. This number (declination of Polaris) is not constant mainly because the **Earth's spin axis gyrates (precesses) similarly to a spinning toy-top on a smooth table,** i.e. the polar distances to stars are changing slightly when the celestial pole gyrates/moves among the stars.

The Sun's declination in the month of July is approximately +23.5 degrees and in January it is -23.5 degrees, and it is zero at the solstices in March and September.

34.2. RIGHT ASCENSIONS

The other necessary celestial (star) coordinate in addition to the declination is called **the right ascension** corresponding in similarity to longitudes on Earth.

Right ascensions are measured along the Celestial Equator *eastward* **from the Vernal Equinox** (where the center of the Sun is around March 21 every year) in units of hours, minutes, seconds and fractions of seconds (0 to 24 hours), or in units of degrees, minutes, seconds and fractions of seconds (0 degrees to 360 degrees). One hour equals 15 degrees (360/24 – 15).

Right Ascensions of 'fixed' stars change 'a little bit' all the time mainly because the First Point of Aries (= one of the intersection points of Earth's equatorial plane with Earth's orbital plane) moves 50 arc-seconds per year because Earth's spin axis (and Earth's equatorial plane) gyrate 360 degrees in approximately 26,000 years like a toy-top. Yes, it is somewhat complicated!

The details of high accuracy declinations and right ascensions are not described in this book. Just be grateful that the specialists take care of those things. Of course, the reader may/can become such an expert herself/himself.

Google: < precession of earth's rotation axis >

As was mentioned earlier, in order **to get a desired star or a celestial object in the crosshairs of a telescope, the object's declination and right ascension must be known. In addition, observing the station's geographical coordinates (latitude and longitude) and the direction to the north (azimuth) are needed together with the sidereal time of the day.** From this data the azimuth (horizontal angle) and the altitude (vertical angle) to the object can be computed. After all this, the telescope can be pointed to the object and the object will not be far from the center of the crosshairs.

Google: **< right ascension and declination >**
                     < noaa geodetic surveying >
                     < celestial sphere >
                     < altitude of celestial pole >
                     < positional astronomy >
                     < terrestrial sphere >
                     < great circle on earth >

http:/aa.usno.navy.mil/publications/docs/AsA_history.html

All star celestial coordinates change very slowly. As was mentioned, **most declination and right ascension changes result from Earth's very slow orbital orientation movements.**

Stellar positions using **Cartesian XYZ coordinates** can also be computed from the equatorial coordinate system of declinations and right ascensions. That spherical trigonometric/mathematical conversion will not be covered in this book.

Google: **< calculating stellar positions >**
                     < cartesian coordinates >

In addition, directions to some stars change relative to the background of more distant stars/galaxies due to so-called **proper motions.** This results from the slow directional changes from the Sun to the star in question. The Milky Way Galaxy is not a solid object. Its stars have their own motions including the Sun.

34.3. PROPER MOTION

Apparent directional motions of stars are called Proper Motions, when their individual declinations and right ascensions change relative to distant galaxies. Most catalogued stars with measurable proper motions are in the same pinwheel arm of the Milky Way Galaxy where we are. If a star is moving directly 'this way' or if it moves directly away from us, then its proper motion is zero, although it may be traveling at a great speed. **All catalogued stars used for astronomical positioning are Milky Way stars.** Stars in other galaxies are completely another matter, and they are not used for astronomical positioning.

    Barnard's Star changes its coordinates faster than all other known stars. Its celestial position changes **only by ten arc seconds per one year**. All other Milky Way stars have more stable celestial coordinates. Declinations and right ascensions of all distant galaxies remain practically constant for eons, even when seen from diametrically opposite points in Earth's orbit. Earth's orbit is small enough to be considered only a point when compared to many light-year distances.

    Apparent visual shapes of naked-eye constellations do not change visibly during our life times. For instance, the "Big Dipper" and Cassiopeia 'W' will look practically the same one hundred years from now as they do today. Some stars with 'large' or uncertain Proper Motions are omitted from accurate observation lists, or the amount of known Proper Motion is taken into account for the time of observation (measurement).

Google: < proper motion of stars >

                < barnard's star >

                < boss general catalog of stars >

34.4. PRECESSION OF EARTH

The precession of Earth's spin axis is mainly caused by the Moon's and Sun's gravitational attractions upon Earth's equatorial bulge in trying to make the bulge (equatorial plane) to coincide with the straight line to themselves. Because the bulge's daily rotational inertia, the bulge wobbles once around in 25,800 years instead of aligning itself with the gravitational pulls of the Moon and Sun.

The gyrating motion of Earth's N-S spin axis is called Precession, where Earth's spin axis describes more or less a circular cone around the pole of the plane of Ecliptic, which pole is 23.5 degrees away from the north celestial pole. North celestial pole is the pole of the Equator (a pole of a plane is 90 degrees up/down from the plane). The declination of the north celestial pole is 90 degrees, and the declination of the pole of the Ecliptic is 90 – 23.5 = 66.5 degrees. The circular cone of precession cuts a small circle on the celestial sphere 23.5 degrees around the pole of the Ecliptic. This gyrating motion is usually much slower (25,800 year-period) than the spinning itself (24 hour-period).

A spinning toy-top on a smooth tabletop gyrates if its spin axis is not vertical and perfectly steady. Earth's spin axis gyrates in a similar manner as the toy-top gyrates. The **gyration (precession) period,** when the spin axis describes a 360-degree conical surface depends on the mass, mass distribution and shape of the spinning body. The spin axis of a spinning toy-top gyrates around the plumb line. The gyration period of **a small toy-top may be a few seconds, but due to the very much larger Earth's mass, the precession period of the Earth is** approximately **25,800 years**.

Prorating the 360-degree **precession motion per one year**, the rate of precession is obtained to be approximately **50 arc seconds per one year** (360 x 60 x 60) / 25,800). This is a major reason why the star declinations and right ascensions change all the time. The drifting (50 seconds per year) Vernal Equinox (first Point of Aries) is the **point of ORIGIN for the Right Ascensions**.

The tilt from the plumb line of the spin axis of a spinning toy-top can vary from zero to 40 degrees or maybe more. **The tilt of the Earth's spin axis is 23.5 degrees from the normal of Earth's orbital plane (66.5 degrees from the plane of the Ecliptic). Earth's equatorial plane and Earth's orbital plane (= plane of the Ecliptic) also form the same 23.5-degree dihedral angle with each other.**

Google: < geodetic precession >
             < precession of earth >
             < precession of rotating earth >
             < precession of earth's rotation axis >
             < earth rotation history >

34.5. THE 23.5 DEGREE TILT OF EARTH'S SPIN AXIS

The tilt of the spin axis may seem to be insignificant, **but this 'little tilting angle' is the only reason for the existence of the seasons of spring, summer, autumn and winter.** The annual seasons are of great importance for life here on Earth.

The 66.6-degree tilt of Earth's spin axis to Earth's orbital plane is the only reason for Earth's seasons: Spring, Summer, Fall and Winter. If Earth's spin axis would be perpendicular to Earth's orbital plane (the Ecliptica), all days and nights of the year would be approximately of equal duration (12 hours), except at the North and South Poles, where the Sun would be on the horizon all the time. For the actual Earth, this situation happens only twice a year, i.e. at Vernal (approximately March 21) and Autumnal (approximately September 21) Equinoxes, when the Sun is directly above Earth's Equator for an instant. At that time the center of the Sun is crossing Earth's equatorial plane. The declination of the center of the Sun is then equal to zero.

The **northern summers are a little longer in duration than the northern winters** because the Earth travels a little faster during northern winters than in northern summers, i.e., the time interval from Vernal Equinox to Autumnal Equinox (northern summer) is a little longer than the time interval from Autumnal Equinox to Vernal Equinox (northern winter). All this is according to Kepler's laws (God's laws really) stating that **planets travel at greater speeds around the Sun when they are nearer to the Sun in their elliptical orbits.**

Google:　　< reasons for the seasons >

　　　　　　< perihelion, aphelion >

　　　　　　< the seasons and the earth's orbit – milankovitch cycles >

　　　　　　< the earth's rotation >

　　　　　　< orbits and the ecliptic plane >

Many of us have noticed that **most globes sold in stores, have their North-South axis tilted by a 23.5 (90 – 66.5 = 23.5) degree-angle** from the plumb line (the vertical), or the spin axis is up from the horizontal plane by 66.5 degrees.

Earth's spin axis is nowadays pointing very close to Polaris, the North Star, and it will do the same again after 25,800 years. Due to precession, after approximately 13,000 years, the star Polaris will be approximately 47 degrees (2 x 23.5 = 47) from the pole of the equator (= the point where the north end of Earth's spin axis is pointing). The pole of the Ecliptic is a much more steady point on the Celestial Sphere than the precessing and nutating pole of the equator. The pole of the Ecliptic is situated at the center of the 23.5-degree (radius) precession circle.

The Vernal Equinox point (= the First Point of Aries) is a point on the celestial sphere, where the intersection line of the Earth's equatorial plane and Earth's orbital plane 'punctures' the celestial sphere. This straight line intersects Earth's orbit at two diametrically opposite points. The other point is the Autumnal Equinox = First Point of Libra.

Google: < celestial coordinate system >
              < equinoxes and solstices usno >

Of course, all this geometry is confusing for those whose expertise does not include dealing with astronomical Declinations and Right Ascensions. The author (just a messenger) believes that the present coordinate system cannot be simplified from what it is.

However, understanding Earth's precession is nothing new. **Already around 130 B.C. Hipparchus knew about the tilt of the Earth's spin axis and even about its precession**, and he was able to estimate the precession period reasonably well.

Google: < astro note 11: celestial coordinate systems >
              < hipparchus >

34.6. NUTATION

The **described precession motion of the Earth's spin axis does not trace a perfectly circular small (23.5 degree) circle on the celestial sphere.**

This wiggling motion is called nutation. Earth's spin axis 'nutates', meaning that the spin axis in its circular precession motion is nodding in and out of the 'perfect' precession circle (23.5 degrees) with an amplitude of approximately 9 seconds of arc having a **main period of 18.6 years. It is caused mainly by the gravitational attractions of the Moon, Sun and by smaller**

amounts by 'nearby' planets upon Earth's equatorial bulge. Recall that Earth's equatorial radius is a little longer than its polar radius. If Earth would be a homogeneous perfect sphere, there would not be any nutations coming from outside gravitational effects.

The length of the 18.6-year period of nutation originates from the nodes of the Moon's orbital plane 'going once around' in 18.6 years. Compare this to Earth's nodes (Vernal and Autumnal equinoxes) 'going once around in 26,000 years'. Earth is more massive object (compare to spinning toy-top) than 'our' Moon.

Google: < earth nutation visualized >
             < wobbling of earth's rotation axis >
             < nodes of the orbit >
             < astronomy answers: planetary phenomena >

34.7. CHANDLER'S WOBBLE

Chandler's Wobble causes small variations in star (and galaxy) Declinations and Right Ascensions. These wobbles move the Earth's North and South Poles (spin axis) around by small amounts. These wobbles are caused by mass movements in the Earth's interior, oceans, atmosphere and by varying snow loads, etc. causing small variations in Earth's moment of inertia. Earth's variable moments of inertia cause the Earth's spin axis to wobble around in an irregular manner. The instantaneous north and south poles move around the mean (average) pole positions by a few meters (yards) with Chandler's period of 435 days. The mean pole is also drifting a little as the graph in the Google site "**the wandering path of chandler's wobble**" shows.

Google: < chandler's wobble of earth >
             < the wandering path of chandler's wobble >
             < wandering path of chandler's wobble >
             < earth's mass, density and moment of inertia >
             < moment of inertia >

For instance, an instantaneous latitude for a point on Earth's surface can vary by 0.5 arc second or 15 meters (50 feet) on the ground during one half of Chandler's cycle (7 months). Recall that the GPS positioning can be much more precise than that. Chandler's Wobble must be taken into account in some precise observations/measurements.

35. Earth's Crust, Mantle and Core in the Balloon Model

Earth's crust is the top-most layer of the Earth's interior. Its average density in most Earth models is 2.67 metric tons per cubic meter or 2.67 times as heavy as water. The thickness of **Earth's crust varies between 20 to 80 kilometers (12 to 50 miles) with locations.** On the room-size balloon model **the thickness of the crust varies from 4 to 16 millimeters (0.16 to 0.63 inches)**, which is comparable to the skin of an apple. The crust is generally thinnest under the deep oceans and thickest under the high mountains.

Under the crust is **the mantle of Earth extending down to almost half way to the center.**

The core occupies the volume from under the mantle to Earth's center. The core has two main parts called the Inner Core and the Outer Core.

The **pressure at the center of Earth is on the order of 3,900,000 times the normal atmospheric pressure at sea level.**

The **temperature at the center of Earth is estimated to be from 8,500 F (degrees of Fahrenheit) to 12,000 F, or from 5,000 K to 7,000 K (degrees of Kelvin).**

The values of **Earth's gravitation and gravity at Earth's center are zero.**

Google: < earth's interior >
            < the interior of the earth >
            < isostasy >
            < isostacy >
            < metric system temperature scales >

197

36. Gravity, Gravitation, Accelerations and Decelerations

Gravitation and gravity are accelerations. The Newtonian gravitational attraction between two masses is called gravitation.

Gravitation on Earth is the pure Newtonian gravitation (attraction), or gravitational acceleration, or pull of gravitation. Gravitation is the vector sum of all the gravitational attractions by every cubic yard (cubic meter) of material throughout the whole body of Earth exerted at any point on the ground, in the air or inside Earth.

Gravity is the vector sum of the Newtonian gravitation and the centrifugal acceleration of Earth's daily rotation, making Earth's gravity slightly smaller than its gravitation.

Gravitation and gravity have directions and intensities; they **are vectors.** The direction of gravity is down in America as well as it is in Australia along the local plumb line towards Earth's center.

Acceleration is increasing the velocity. Deceleration is decreasing the velocity (braking). Compare this to driving a car. When the speedometer number is increasing, the car is accelerating; when braking, the speedometer number is decreasing.

The units of accelerations are: centimeters per second per second = centimeters per seconds squared or feet per second per second = feet per seconds squared.

Keep in mind that gravitation and gravity are accelerations. Gravitation is a constant, mutual and unrelenting pull of attraction between all masses (compare to two neighboring magnets). Gravitation on Earth describes how fast a freely falling body is accelerating (= increasing its speed) when it is falling freely toward the ground (ignoring air resistance). For instance, a freely falling mass on Earth is increasing its speed (ignoring air resistance) by approximately 10 meters (1 G = 9.81 meters per every falling second = 33 feet per every falling second) during its free fall. For example, after a 5-second free fall, the falling mass (ignoring air resistance) has attained a speed of 5 x 9.81 = 49 m/s = 161 ft/s = 110 MPH = 176 km/h. The distance of that 5-second fall is 123 meters = 402 feet.

Bombs dropped from airplanes are good examples of this type of situation.

Accelerating a car on a horizontal surface provides another example on accelerations. For instance, making a car go from a standstill to 60 MPH = 96.6 km/h in 4 seconds on a horizontal surface, the average horizontal acceleration of the car and driver is = 1.21 G. If a 150-pound (= 68-kilogram) driver would be sitting on a special tilting bathroom scale during those 4 seconds (the vector sum of his/her normal weight plus his/her acceleration), the scale would indicate his/her weight to be 182 lbs = 82 kilograms during those 4 seconds.

Deceleration means reduction of speed, such as in braking a speeding car. Objects that are changing their speed or their direction of travel are also said to be accelerating/decelerating, such as making a speeding car to make a sharp curve on level road when loose objects in the car try to slide sideways by the force of **centrifugal acceleration.** Like gravity, the rate at which the direction of motion changes is also measured in units of meters or feet per seconds squared.

When speeds are being reduced (such as when braking a car), the process is called **deceleration.** If a car traveling at 110 MPH = 177 km/h is braked to a standstill in 5 seconds, the average deceleration is = 49.1 m/s/s = 161 ft/s/s = 5 G. A 150-pound (= 68 kilogram) passenger would press the dashboard as hard as a 750-pound (= 340 kilogram) mass presses the ground. Note again that 5 x 9.81 m/s/s = 49 m/s/s = 5 x 33 ft/s/s= 161 ft/s/s, and that with normal tires on normal surfaces, this braking of 5 G cannot be done in 5 seconds.

When moving objects change their directions, such as a passenger in a car making a sharp left turn, they are pushed towards her/his door.

Earth's daily spinning reduces (opposes) the gravitational pull more so near the equator than near the poles. The daily spinning of Earth produces a small centrifugal acceleration reducing our weights by up to 0.3 % near the equator than near the poles. **Combination (vector sum) of gravitational pull and centrifugal acceleration is called gravity.** Gravity can be compared to accelerations experienced on amusement park carousels having inner seats (closer to the spin axis) for small children and the outer seats for the more daring.

Some examples of **centripetal acceleration** are:

1. Our Moon orbits the Earth at a distance of approximately 60 times the Earth's radius. Earth's gravitational pull forces the Moon to orbit the Earth instead of letting it fly tangentially along a straight line away from Earth.

2. The seat supporting chains of amusement park carousels force the seat to move in a circle instead of letting the seat fly in the tangential direction.

Some amusement park rides (such as roller coasters) are characterized by rapid changes in speed and or direction. These rides have large accelerations. Rides such as the carousel result in relatively small accelerations, the speed and the direction of the riders change gradually (less suddenly). Fast carousels are centrifuges.

Google: **< gravitation by Isaac Newton >**
            < physics in the amusement parks >

The seat-supporting chains on amusement park merry-go-rounds force the chairs from their forward straight motion to curve the travel into a circle. The horizontal component of the stretching of the chains is an example of **centripetal acceleration** pulling the chairs toward the center of rotation. Another example of **centripetal acceleration** is provided by Earth's gravitational pull on our Moon forcing it to orbit the Earth, instead of flying straight at its speed of 1 km/s = 0.6 miles/second into its oblivion in the emptiness of surrounding space. The magnitude of Earth's gravitational attraction on the Moon is 0.0027 G.

Google: **< gravitation and gravity >**

Each one of us has been dealing with gravity in some way every instant of our lives. The balance we try to keep, every step we take, every object we move (even a finger), sitting in a chair and moving our bodies around is affected by the pull of Earth's gravity. Gravity 'sucks' us all the time toward the Earth's center whether one is on the ground, in a submarine, on a boat, in an airplane or in the orbiting Shuttle. Without a support capable of stopping (centrifugal acceleration for a

Shuttle) our fall toward the Earth's center, we will keep falling to lower altitudes, through quicksand and through trap doors, etc.

Gravity has its effects on us even when we sleep. When sleeping, after so many minutes/hours, most people move/rotate themselves so that the pressure caused by the pull of gravity on our skin against the mattress does not press the same part of the skin all night long. Lying in one position for a long time may cause painful bedsores. This is a problem for long-term bed patients.

Gravity keeps the ocean waters pressing the ocean floors on the American shores as well as on the Australian shores. Gravity keeps sea level in a nice balance all around the Earth. Gravity makes the river waters flow. Gravitation and gravity are everywhere on 'our' spinning Earth. Outside this spinning Earth, there is only gravitation.

Only the Earth-orbiting astronauts and the Moon astronauts have experienced almost weightless conditions for *extended periods* of time. There are many adjustments to make in normal daily living when in weightless conditions, where the pull of (Earth's) gravity is not felt. Even the orbiting astronauts/cosmonauts are pulled 'down' by gravity toward Earth, but that pull of gravity is counter-balanced by the centrifugal acceleration of the orbiters. All of us, either consciously or unconsciously, are influenced by gravity, and we deal with it in some manner all the time.

Our reactions to gravity for our balance are sensed/dealt with our more or less automatic sensors in our bodies (ears, eyes) and by our muscles. Our abilities to walk/run and the ability to stand upright are just a few examples of our responses to gravity. As our bodies move, the fluid in our inner ears moves around by the force of gravity providing our brains with information on our head/body orientation. That part of the human ear serves us as an accelerometer for our body- orientations. Orbiting astronauts cannot feel where 'up' or 'down' is in their Shuttle due to their *almost* zero-gravity conditions.

Google: < ear clinic information center – balance system 101 >

It is the gravity that gives us (skinny, normal or obese) **our body weight** proportional to our body mass. Without gravity we would be weightless. Gravity pulls us toward the ground 'like a magnet' all the time, everywhere and anywhere. All masses have gravity fields.

When a passenger is sitting in an airplane seat and the pilot changes the direction of the plane by tilting the plane correctly, flying 'perfectly' at a constant altitude/speed, the passengers usually do not feel the change in gravity, although the centrifugal acceleration of the turn causes a slightly higher pressure on their airplane seats. During the turn passengers are pulled 'down' against the seat by the sum of Earth's gravity and airplane accelerations (centrifugal acceleration of the turning airplane, airplane accelerating up or down). If a passenger would be sitting on a special tilting bathroom scale in her/his seat, the scale would indicate a 'heavier' weight in the turn than when flying straight at a constant altitude at constant speed. When fighter pilots make tight turns at high speeds, their seats are experiencing weight increases by several times their weight.

Earth's gravitation is also strong enough to hold our atmosphere with the fast moving orbiting Earth so that there are oxygen molecules for us to breathe and carbon dioxide for the green growth here on Earth. The Moon's gravity is not strong enough to keep gas molecules down to the Moon's ground.

37. Moon's Gravity

The average distance to the center of the moon from Earth's center is sixty times Earth's radius or 1.2 light-seconds. The intensity of the pull of **Earth's gravitation keeping the Moon in its orbit around us is only one part in 3,600 (3,600 = 60 x 60) of one G, or 270 milligals or 0.000270 of one G.**

Recall that Earth's gravity is = 1 G = 981 centimeters per seconds squared = 981 gal = 981,000 milligals = 32.2 feet per seconds-squared.

Also recall that Moon's gravitation on its surface is 165,000 milligals or approximately one sixth of Earth's gravity (981,000 milligals) at sea level. The Moon's centrifugal acceleration is very small because it spins around its axis only approximately once per one 'moon' (month). Therefore the **Moon's gravity is almost entirely the same as its pure gravitation.**

The Moon's gravity is not strong enough to keep an atmosphere. Any gas on the Moon will escape fairly quickly to the surrounding space just because its gravitation is not strong enough to hold the freely-moving gas molecules to itself.

Where will those escaped gas molecules go from there? The Moon's gravity cannot hold them near to its ground. Those gas molecules will still be pulled/influenced by the gravitations of the Sun, Earth, Moon itself and also by the Sun's radiation pressure trying to push them away from the Sun. The Sun's average gravitational pull in the Moon's neighborhood is 590 mgal = 0.590 gal (gal = centimeters per second squared) or 0.0006 times one G. (One G = 981,000 mgal is Earth's normal gravitation at Earth's surface). Earth's gravitational pull in the Moon's neighborhood is 270 mgal and Moon's own gravitation at Moon's surface is 160 mgal. Therefore, near a new moon the escaped gas molecules will start to 'fall' toward the Sun, and near a full moon they will start to 'fall' toward the Earth and Sun. **Those molecules will become orbiters in this solar system.** They will surely get spread around!

Google: < solar radiation pressure >

Compared to a speeding car, where the air pressure is greater against the windshield than against the rear window, Earth's high orbital speed does not cause (to the knowledge of the author) a higher air pressure against the front side of Earth than against the rear side of Earth as it is speeding in the vacuum of space at 30 kilometers (20 miles) per second in its orbit while spinning/rotating at the same time 360 degrees per day.

Google: < gravity on the moon >
< escape velocities >
< radiation pressure by the sun >

We on Earth have been blessed with an atmosphere so close to ideal, that very few people in the medical profession or in earth sciences can even imagine how to improve it even if the 'all powerful magic wand' would be available. Are we grateful to God/Creator, or to whomever we think made the atmosphere for all of us? **We need/want to breathe every minute of our lives, don't we!** This Earth is the only planet/moon in the known universe with breathable air. There might be other earthlike planets out there, but so far nobody knows anything about them.

Google: < terrestrial escape velocities > Scroll down to see the situation for Moon and planet Mercury.

38. Sun's Gravity

Gravity on 'our' Sun's surface is approximately 28 G. (One G = average gravity on Earth = 9.81 meters/second/second = 32 feet/second/second). At its equator, the Sun's surface rotates once around in about one month. The length of the Sun's radius is 1.8 times the distance from here to the Moon. The Sun's centrifugal acceleration at its equator is approximately 408 milligals = 0.0004 G. The small value of 0.0004 G compared to Sun's 28 G gravity does not fling any masses off the Sun's equator into the outer space.

As was mentioned, the Sun rotates on its axis once in approximately 27 days. This rotation rate varies with the Sun's latitude since the Sun is a ball of hot gas. Its equatorial regions rotate faster than its polar regions. The equatorial regions rotate once around in approximately 25.6 days, and its polar regions rotate once around in about 36 days. The Sun rotates in the same direction as Earth (eastward). The Sun's equator is tilted by about 7.25 degrees to Earth's orbital plane (Ecliptica).

Google: < solar rotation >
          < solar gravity >
          < soho >

The Sun's gravitational pull (attraction) keeps this Earth in its annual orbit around the Sun so that the Earth does not fly tangentially away from 'our' life sustaining Sun into a super cold oblivion. **Similarly, 'our' Sun's gravitational pull keeps all other solar planets with their moons, the asteroids and comets in their orbits.** The numerical value of the Sun's gravitational attraction (pull) here at one AU = 149,600,000 kilometer = 93,000,000 mile distance is approximately 590 milligals or 0.0006 of one G. That is all that is needed to keep this Earth in its annual solar orbit with its 30 kilometers (19 miles) per second speed.

Gravitation and gravity are vector quantities (keep reading, there will be no vector calculus in this book!), meaning that they have certain numerical values of magnitude or intensity. In addition, they have their own directions, as Isaac Newton said.

To properly describe the force of gravitation/gravity affecting any object, one must specify the direction and mass (the size of the object in kilograms or pounds), which is under the influence of the prevailing gravitation/gravity.

For example, consider a 150-pound (68 kilogram) person standing next to a railroad locomotive. Both are then affected by approximately the same gravity (gravity on Earth varies slightly with elevation and location). The force pulling the standing person against the ground under her/his feet is only a tiny fraction of the force of the pull of gravity of the locomotive against the tracks, ties and the ground, remembering that the gravities affecting the person and the locomotive are the same.

The all-important and simple formula from Physics states this fact: the force is equal to the mass times the acceleration. The direction of the force (pull) of gravity is the same as the direction of the acceleration, or on Earth, it is along the plumb line straight down everywhere and anywhere on Earth as well in America as in Australia.

Consider another example. Lifting a 150-pound (68 kilogram) person on the Moon would take as much effort as lifting a 25-pound (11 kilogram) child here on Earth. Imagine lifting (ignore also the temperatures) a 150-pound (68 kilogram) person on the Sun's surface would take as much effort as lifting 4,200 pounds (1.9 metric tons) here on Earth. Note that the mass (150 pounds) in these examples is the same on the Earth, Moon and Sun, but **prevailing accelerations of gravity surely make the forces very different**. By the way, the Sun's gravitation, approximately 27 G, is so strong that nobody could do anything, even lift a finger under such a strong crushing gravitation. This can be checked out in a centrifuge, where any person will lose her/his consciousness already at around 10 G.

High accelerations can be duplicated in centrifuges, where 10 G accelerations pretty much flatten a person against the outer wall producing unconsciousness or death fairly soon. Healthy persons can tolerate 10 G accelerations and decelerations for a short time. Imagine what would happen to the human brain, heart, bones, lungs, eyeballs, fluid in the inner ear, anything in the stomach, etc. under

10 G, or higher accelerations! To help such imaginations, think of 'replacing them' with equal-size pieces of lead ignoring the body-functions. If the eyeballs were that heavy they could come halfway out of their sockets.

High **decelerations** have similar effects as high accelerations on human bodies. Car accidents and airplane crashes produce deadly decelerations for short periods of time every day with tragic results.

Some circuses have cannons for shooting 'human torpedoes'. They provide good examples of rather large short-term accelerations and decelerations. In the cannon tube, the 'human torpedo' experiences a sudden acceleration and at the end of the flight comes the deceleration.

Like accelerations, velocities (speeds) are also vectors. To fully describe the meaning of a certain velocity, it is necessary to state how fast and in which direction the motion is. It makes a difference whether one is moving toward something or away from it, as in coming and going.

39. Examples of High Accelerations

The intensity of the pull of gravity here on Earth is six times stronger than the pull of gravity on the surface of the Moon, and the pull of gravity on the Sun's surface is approximately 28 times stronger than on Earth.

Gravity can be increased in centrifuges, but the persistent unrelenting pull of gravitation cannot be canceled. Even the astronauts in their Shuttles are subject to Earth's gravitational pull. The Earth's gravitation pulls them toward the Earth. At the same time their orbital velocity of some 7 kilometers (4.3 miles) per second produces a centrifugal acceleration away from the Earth with the same intensity. So, every atom/molecule/cell in the astronaut's bodies are pulled in two opposite directions by equal forces of acceleration. Humans cannot feel such 'stretching' of their atoms/molecules/cells in their bodies.

Some US Air Force/Navy/Marine pilots are tested for their tolerance of accelerations in special centrifuges. Young adults can tolerate accelerations (G-forces) up to 10 G for short periods of time before losing their consciousness.

A pilot, or any person being tested, sits in a chair facing the center (spin axis) of the centrifuge with his/her back toward the centrifuge's outer wall. The 10 G-force (centrifugal acceleration) will push their bodies against their seatbacks with a force ten times their bodyweight. Centrifugal acceleration will pull their facial skin and all body tissues around their bodies toward their backs. The corners of their mouths move toward their ears. Their heads will slump sideways because the centrifugal acceleration tries to pull the center of mass ('center of gravity') of their heads as close as possible in the direction of the centrifugal acceleration, i.e. as close as possible to the supporting seatback. Compare this to a rock trying to roll downhill. The subject's breathing becomes difficult and their heartbeats are bothered among many other bodily discomforts.

If a catapult on an aircraft carrier launches an airplane to a speed of 180 MPH (290 km/hour) in 2.5 seconds, the pilot will be under an average acceleration of 3.2 G for those 2.5 seconds.

In addition to their centrifuge, located at Holloman AFB (Air Force Base) in New Mexico, a 10-mile long High Speed Test Track (precisely surveyed to be straight) has provided accelerations (speed increases) and decelerations (braking, speed reductions) up to 200 G's for short periods of time. A few pilots have taken rides on the Test Track enduring strong accelerations and decelerations up to 40 G's **for very short periods of time.** On Dec. 10, 1954, Dr. John Paul Stapp rode a jet engine powered test-track sled to a record speed of 632 MPH = 927 ft/s =1017 km/h = 282 m/s, decelerating (braking) to zero speed in 1.25 seconds with an average deceleration of approximately 35 G. For those 1.25 seconds his average body weight was about 6,800 pounds (3.1 metric tons). Decelerations put similar strains on a human body as accelerations do. Strong accelerations are 'sudden and fast departures'. Strong decelerations are accomplished by heavy braking effectively reducing high speeds very quickly.

Google: < deceleration project of Paul Stapp >

Many examples of high G-forces are described almost daily in newspapers. As an example, consider a car going at a speed of **67 MPH = 108 km/h = 30 meters per second = 98.4 feet per second and hitting at this speed an unyielding rock wall at a 90 degree angle.** Assume that the driver is protected by a 'perfect' whole body-fitting airbag, which is able to stop her/his forward motion in one-meter (39 inch) distance. During a uniform deceleration over that 3.28-foot (one-meter) distance, the driver's deceleration (reduction of speed) would be 46 G for 0.067 seconds, **which would be lethal for most people in any car.** Now you know how not to drive!

If in the same situation, the same driver's imaginary very large (6.56 ft = 2 m) airbag system would bring the driver to a stop with a uniform deceleration from the 67 MPH speed to a stop in 6.56 feet (two meters), the experienced deceleration would be 23 G's for 0.133 seconds. The driver may or may not survive that deceleration. An airbag that large would fill the whole front seat.

The deceleration over the two-meter (6.56 feet) distance alone imparts a bone-breaking force of 23 G on the human body. **To a 150-pound (68 kilogram) person, the felt pressure is similar to being under 3450 pound (1550 kilogram) load for 0.133 seconds. It is a very unfriendly extra jolt of weight to endure.** In addition, in such crashes, there are many opportunities/dangers of being bruised and cut by objects, which are not as soft as the airbag.

In actual car crashes the decelerations are not quite uniform because the car parts being crushed are not of uniform strength. The given deceleration values are good averages, although in actual car crashes, greater and smaller decelerations will be experienced during the period of coming to a stop.

May these examples make us all safer and wiser drivers!

Similar great accelerations are experienced by some poor souls in some explosions, and then after the flight, the decelerations when coming to a more or less sudden stop. The sudden acceleration comes from the sudden gas pressure blowing away from the explosion together with some debris. Sudden decelerations are just as deadly as sudden accelerations of similar G-values.

The Creator made us to be comfortable under one-G-acceleration conditions. We experience one G acceleration all the time with only a few exceptions of short durations. Exceptions to situations differing from one-G-situations are provided by astronauts, fighter pilots, acrobatic pilots, racecar and some other car drivers, amusement park riders and a few people in some special airplanes, etc., who experience accelerations differing by appreciable amounts from zero G to a few G's.

Google: < what is the acceleration of gravity >
            < vectors >

40. Back to Earth's Gravity

It is commonly assumed that the intensity of Earth's gravity is 1 G (9.81 meters per second squared = 32.2 feet per second squared) everywhere on Earth, but 'there is much more to it'. **One G is the average value of gravity at the sea level on Earth's surface. Earth's gravity is the vector sum of the Newtonian gravitational attraction/pull of the total body of Earth plus the centrifugal acceleration coming from the daily spinning of Earth and of the tidal gravitational attractions by the Moon, Sun and 'nearby' planets.**

Also, the gravity is a function of topographic elevation and of local mass irregularities within the body of Earth.

Google: < gravity anomalies >

Gravitation describes the existing mutual pull between two masses in question. If this Earth would not rotate, it would have only gravitation. Rotating Earth has gravity. Earth's gravity is pulling lake waters, trees and each and everyone of us as well as our belongings such as pens, pencils and papers, in the direction to Earth's center whether we are in America or Australia. Each one of us is also simultaneously pulling Earth toward ourselves.

According to Newton's Law on Gravitation, the gravitational attraction is proportional to the masses involved; therefore, when we stop supporting (drop) a pencil, it falls to the floor/ground/etc., but the floor does not jump to the pencil.

As we know, Earth is very closely a spherical globe. Sometimes one hears that the Australians are living 'down under'. **We all also are living 'down under',** when compared to somebody else's horizon that is 180 degrees, or 20,000 kilometers (12,000 miles) away from us along Earth's curved surface. **Or, we are living on the 'vertical sides' of the globe** when compared to somebody else's horizon 90 degrees, or 10,000 kilometers (6,000 miles) away from us along Earth's curved surface. Earth's gravity is pulling all masses, i.e. all animals, cars, trucks, trains, etc. against the ground. **When walking, riding a bike, or driving a car anywhere on Earth, it is OK to consider that one is under this Earth or 'clinging' to its side, or being on top of it.** The emptiness of space is all around this small precious round planet.

40.1. EARTH'S GRAVITY AND CENTRIFUGAL ACCELERATION

Earth's centrifugal acceleration (brought about by daily spinning of Earth) **is at its maximum at high topographic elevations at/near Earth's Equator where it reduces the gravitational pull by approximately 0.3 percent, thus making the gravity to be** approximately **0.97 x G at those places. At high latitudes, the gravity is slightly greater than the average value of 1 G = 9.81 meters per second squared = 32.2 feet per second squared. At the poles, the sea level value of gravity is about 0.2 % greater than 1 G.** Note that at the South Pole, the topographic elevation on the ice sheet is about 3 kilometers (2 miles).

Google: **< international gravity formula >**
             < world geodetic system 1984 – background >

The small variations in gravity have an **appreciable effect on Olympic and World Records in many sports events.** Among these are: weight lifting, high and long jumping, pole vaulting, javelin throw and shot put. Therefore, it really depends where on Earth an Olympic/World record was/is made. **The actual value of gravity could be easily measured at the place where the record was/ is made.** From the known value of gravity, the correction to the achieved record can easily be computed to one fair world standard (one G). For instance, **in the javelin throw, the correction can be several inches (centimeters). It remains to be seen when the World/Olympic record keepers will become sufficiently sophisticated to be fair to the best athletes.**

   The variable tidal pulls by the Moon and Sun on all earthly material cause relatively small variations in Earth's gravity, but the pulls have sizable effects.

Those objects (bodies, masses), **which are on the move** (cars, trains, airplanes, artillery shells, etc.) **with respect to the Earth's surface, will have their own speed-dependent centrifugal and/or Coriolis accelerations.** For instance, a car traveling in a westerly direction presses the road surface a little harder than it presses the same road surface traveling in an easterly direction at the same speed (assume the same mass for the car). When traveling in a westerly direction, a part of the centrifugal acceleration (coming from Earth's spinning eastwards) is cancelled, and when traveling in an easterly direction, the car's speed adds to the centrifugal acceleration, which decreases the pull of gravity of the car.

Google: < coriolis force >

Most people have experienced/observed centrifugal accelerations, for instance on merry-go-rounds, in cars changing direction of travel at high speeds and from water droplets flying off bicycle tires. Human bodies cannot feel Coriolis forces.

40.2. EARTH'S GRAVITATIONAL PULL HAS DEFORMED 'OUR' MOON

Earth's gravitational pull on the body of 'our' Moon has caused strong tidal effects there. The Moon has no liquid water to pile up for a tidal wave, nor has it any mean sea level for a reference surface for its topographical elevations. As a result, the Moon's ground now has a stationary 'frozen' tidal hump (wave, bulge, systematic high ground) along the line from Earth to the Moon. In the process the Moon's rotundity and also its spin (rotation) rate have been 'messed up' rather permanently.

The Moon's average equatorial radius is 1,737.4 kilometers (1,079.6 miles). **Earth's gravitational pull has 'stretched' the Moon's diameter in the direction toward Earth to be longer than its 'crosswise' diameter. The Moon is now a slightly out-of-round spheroid resembling an egg with its longest diameter pointing toward the Earth** *all the time* **with only small librations, (wiggles, wobbles).** For this reason, the Moon's center of gravity (the Moon's spin axis goes through that point) is approximately 2 kilometers (1.2 miles) 'this way' from the halfway point of its longest diameter pointing to Earth. Therefore, the Moon's monthly spinning is wobbly (like a thrown spinning carpenter's hammer on slippery ice) and not like a smooth spinning sphere.

Therefore, the same Moon's hemisphere with small wiggles (lunation, librations, up/down, left/right) is always pointing 'this way'. Due to these librations, a little more than one half of the Moon's surface is visible from Earth during a lunar month. The longest Moon's diameter is not pointing **steadily/exactly** to the Earth's center, but it 'wiggles'/oscillates by small amounts to the 'left and right', 'up and down' so that a little more than one half of the Moon's surface can be seen from Earth. **This phenomenon is called the Moon's libration.**

In addition to the large tidal effects on 'our' Moon caused by Earth, the Sun is also causing some (much smaller) librations approximately up to 0.04 degrees. The eccentricities of Earth's and Moon's orbits are also slightly perturbing the Moon's librations in both east-west (left/right) and north-south (up/down) directions.

Google: < **librations of the moon** >

              < **APOD: November 8, 1999 – Lunation** > This site has a good video.

              < **the moon** >

              < **astronomy picture of the day archive** > Click on 2005 November 13:

http://antwrp.gsfc.nasa.gov/apod/ap051113.html
http://science.nasa.gov/headlines/y2005/07dec_moonstorms.htm?list186835

As was mentioned, in addition to the stretching of the Moon, Earth's tidal effect has been (and is) so strong that **it has forced the Moon's rotation (spin) rate to be the same as its monthly rate of revolution around this Earth.** Thus the Moon's spin rate and its revolution rate are in lockstep or synchronized, each going once around (spinning/rotating, revolving) during the months (moons).

The Moon's far side/backside cannot be seen from Earth.

Google: < **diameter of the moon** >

              < **tidal effects of moon and sun** >

< tidal forces >
< newton's law of gravitation >
< latitude dependent changes in gravitational acceleration >
< moon fact sheet >

40.3. DIRECTION OF GRAVITY, ZENITH (= UP) AND NADIR (= DOWN)

There is no absolute universal direction 'up' or 'down' from Earth to the skies except for one spot at a time. Every radius emanating from the center of the spherical Earth is very close to the plumb line at the point in question. The directional pull of gravity of the (almost) spherical Earth determines the 'ups' and 'downs'.

The point straight up along our plumb line toward the sky is our zenith for any and every location on Earth. It is at the extension of the plumb line straight up from that location to the celestial sphere 90 degrees above the horizon. As Earth turns daily once around its North-South spin axis, the zenith point of the location describes a diurnal circle on the celestial sphere among the visible 'nearby' naked-eye stars in the sky, whose declination (angular distance from Celestial Equator) is the same as the latitude of the point in question. For instance, at 40 degrees North latitude, the declination of the zenith is 40 degrees north, and the diurnal circle has a polar distance of 50 degrees around the North Celestial Pole.

The nadir point is 90 degrees below the horizon under our feet on the celestial sphere. Zenith and nadir points of one spot on Earth are diametrically opposite points on the celestial sphere. They are also antipodes to each other, and they are the two poles of their common horizontal plane. Zenith and Nadir points are called the poles of the horizontal plane. Earth's North and South Poles are the poles of Earth's equator and also the poles of the horizontal planes at North and South poles.

On the scale of the celestial sphere, Earth's equatorial plane and the horizontal planes on Earth's North and South poles are all the same plane, because the miniscule separation of these parallel three earthly planes can be considered to be equal to zero. Compare Earth's diameter, which is less than one light-second to the infinitely large radius of the celestial sphere, or even to the naked-eye celestial sphere with its radius of 2,000 light-years = 63,000,000,000 light-seconds!

On Earth's surface two diametrically opposite points are also called antipodes to each other. For instance, on Earth the South Pole in the Antarctica is the antipode of the North Pole in the Arctic Ocean and vice versa. New Zealand is near the Antipode of Spain. The Antipode of 40 N; 105 W (not far from Denver, Colorado) is at 40 S, 75 E in the middle of the Southern Indian Ocean.

Because each point on Earth has its own zenith and nadir points and due to the spheroidal (= almost spherical) shape of this planet, the zenith and nadir points cover the whole surface of the celestial sphere, when the whole Earth is considered. They are all over the heavens. For instance, the Zenith points of two locations, which are 69 miles (111 kilometers) apart, are one arc degree apart on the celestial sphere. For comparison of angular 'sizes', the visual diameters of 'our' Moon and Sun are approximately one half degree.

The straight-line distance through Earth's center from any point to its antipode is approximately twice the mean radius, or 2 x 6371 = 12,742 kilometers (7,917 miles). The distance along the spherical surface between a point and its antipode is approximately 20,000 kilometers (12,430 miles). On the celestial sphere the distance between a zenith and its antipode is 180 degrees.

If we are stationary on Earth, our zenith point in the sky sweeps through every point of its diurnal circle during every 24 hours. Therefore, the direction 'straight up' from each point on Earth varies among the stars as Earth rotates. For 40-degree north latitude, the location of the zenith among the stars is 100 degrees away from what it was 12 hours earlier.

Google: < celestial star coordinates >

It would not be wrong to think sometimes when walking/driving/riding that Earth is above you instead of under and that you are the one under the globe, or that the globe is on your left or right side, as some aerobatic pilots might see 'things' in their situations. It is the gravity of Earth that is 'sucking' (pulling) you (and your car) making you to cling to Earth and press the road surface with your total weight whether you consider to be on top, under or on the sides of Earth.

Consider a globe depicting Earth or even a basketball as a reasonable three-dimensional model. **Any radius can be the 'up' direction (vertical) at the location, where a radius comes out of the globe. Wherever one is on Earth, one has her/his own unique plumb line vertical direction, which our brains can determine using several sensors in our bodies to sense the direction of gravity helping us to keep our balance.** When America is on top of the globe, the horizontal planes in the Mediterranean seem to be vertical, the Nile River seems to be flowing up in its northerly direction and the whole Mediterranean seems to be ready to spill its waters over Syria and Israel. Other places on Earth have their own plumb lines and horizontal planes determined by Earth's gravity.

Google: < inner ear, balance 101 >

It is the gravity that keeps the rivers flowing and the ocean waters in a nice, almost spherical global shape. Yes, Virginia, the Earth is round!

40.4. EARTH IS ROUND DUE TO ITS GRAVITATION

It is the gravity that makes this Earth round. Gravity/rain/wind continuously makes the Earth rounder by transporting some masses to lower elevations. The Sun, Moon and eight (other) solar planets and their 'larger' moons are also essentially round due to their own gravities.

The Moons of the planet Mars, Deimos with an average diameter =12.6 kilometers = 7.8 miles and Phobos with an average diameter = 22.2 kilometers = 13.8 miles are irregular rather than spherical. Saturn's moon Mimas with an average radius = 397 kilometers =247 miles is almost spherical, but Hyperion with an average radius = 145 kilometers = 90 miles is not spherical. The gravity of Mimas is strong enough to have made it spherical.

Google: < the moons of mars >
          < the moons of Saturn >

The Earth's inertia keeps it rotating around its N-S spin axis, and inertia also keeps the Earth orbiting the Sun in the gravitational field of the Sun.

Google: **< inertia and mass >**

As has been mentioned, we have sensors in our bodies so that a healthy person unconsciously feels and knows the approximate direction of gravity, or where 'up' or 'down' is, whether in America or in Australia. Without our working sensors we cannot stand up without falling/collapsing. With acute severe inner ear problems, one may be able to crawl but not walk.

Google: **< human inner ear and balance sensors >**

One could compare our walking on Earth's surface to ants or houseflies, which can walk over, under and on the sides of a model globe sitting on a desk. (OK, get the fly swatter!) The pull of gravity is not the dominant force with the walks of ants and houseflies. Those insects have sticky feet or suction cups enabling them to seemingly defy gravity. It is the gravitation (gravity for rotating bodies) that makes us cling to Earth anywhere and everywhere all around this Earth.

Where the direction 'up' (straight overhead to the zenith) is at one location in the United Sates of America is not 'up' in Australia, Japan or Europe. The straight 'up' direction on the floating sea ice at the North Pole is almost 180 degrees different from the 'up' direction of the floating ice-shelves around the Antarctica.

We have a tendency to believe that we are on 'top of the world'. This round Earth curves down from us in all compass directions. This applies to any and all locations on Earth.

Due to the many mass irregularities inside this Earth, the plumb lines on Earth are slightly curved and not exactly straight lines going to the center of the Earth. A hanging plumb bob is a tangent to the slightly curved plumb line, and it determines the 'upright' direction at any location on Earth.

The mass irregularities inside the Earth also make the mean (average) sea level (the so-called Geoid) have some rather permanent undulations (bumps) of the order of magnitude up to plus-minus 150 meters (yards). These bumps are small 0.002 % variations in Earth's rotundity.

The directions of plumb lines (perpendicular to level surfaces, mean sea level, i.e. normal floors, shelves, etc.) differ in California and New York approximately by 45 degrees and so do the mean sea levels on the US east and west coasts.

This is obvious to all when looking at a model globe, but for 'everyday-living', most of us have the 'Flat-Earth-Society' mentality considering only the tangent plane within our horizon. Even when we travel/move around, we are always at the center of our instantaneous horizon.

When a tennis ball is thrown almost straight up anywhere on Earth, one can (usually) catch it a few moments later. The tennis ball follows the rule that 'What goes up, must come down'. On the other hand, some optimistic gardeners have a saying: 'What goes down, must come up'!

Going back to the tennis ball and omitting the air resistance **during its flight, the tennis ball is in a Keplerian elliptical orbit around the center of gravity of the Earth.** Only a small portion of that orbit ellipse (trajectory of flight) is above Earth's surface. This is essentially the same situation when a comet orbits the Sun or when an artificial satellite orbits the Earth.

One can imagine a situation where the whole Earth's mass (including the atmosphere) would be concentrated at one point at Earth's central point. Then a thrown tennis ball (here at 6371 kilometers from the central point mass) would be an Earth-satellite if it would not be stopped nor retarded on its way down. That tennis ball would fall some 6371 kilometers (3959 miles), increasing its velocity until its orbital pericenter (nearest orbital point to the central point-mass of Earth). After that pericenter passage, it would 'complete its swing around' the central point-mass in its Keplerian orbit starting to slow down from its maximum orbital speed climbing up and continuing to slow down to the same velocity as it was originally thrown at the same 6371-kilometer distance from the central mass-point. It would go just as high as it was originally thrown, and at this maximum altitude it would start falling down again repeating another similar Keplerian orbit around the Earth's central point mass.

Artillery shells and bullets on their way to their targets are also in elliptical orbits around Earth when disturbing effects, such as the air resistance/wind are omitted. **In the 'olden days', the artillery shell trajectory computations were often made, assuming that the trajectory was a parabola as a good enough approximation to the actual orbit ellipse.** Computations using parabolas were a little easier than using ellipses according to Kepler's laws governing flying (orbiting) objects through a gravitational field of another object. Nowadays when good computers are readily available, it does not much matter whether an ellipse or a parabola is used in such computations. According to Kepler's laws ellipses are mathematically correct for this purpose. Parabola usage brings just an unnecessary approximation into the results.

Many comets are in highly eccentric elliptical orbits around 'our' Sun. For instance, the nucleus of Halley's comet in its 76-year orbit will be at its aphelion in December 2023, approximately at 35 Astronomical Unit distance from the Sun in its Keplerian orbit. (35 AU = 4.86 light-hours.) It could be said that, at the aphelion point, the most distant orbital point from the Sun, the comet's altitude from the Sun is at its maximum.

Google: < geoid undulations >
             < ocean surface topography from space >
             < comet halley >

41. WORLD ATHLETIC RECORDS AND GRAVITY

Some of us remember the Moon-astronaut, Alan Shepard Jr., and the golf balls he hit on the Moon. When his work at hand on the Moon was finished, Shepard pulled out two golf balls and unfolded a collapsible golf club made specially for the occasion. Due to his spacesuit, he had to swing the club with only one hand. He became the first person ever to hit golf balls on the Moon. **Had the golf balls been hit with the/in same force/direction here on Earth, they would have flown a much shorter distance than they did on the Moon.**

Google: < alan shepard >

The unequal pulls of gravities on the Earth and Moon make a big difference on how far thrown objects (golf balls, Olympic javelins, etc.) **fly.** Gravity on Earth is approximately six times stronger than on the Moon. Earth's gravity is not quite constant everywhere on Earth. It depends on the topographic elevation, latitude and on a few other items. **Variations in the gravity do have a sizeable effect on many Olympic and world athletic records.** The intensity of gravity can easily be measured using gravimeters with sufficient accuracy at any competition place, and the record-making results can be easily computed for one agreed standard value of gravity right on the spot before the result is given.

Google: < mass and spring measurements >
             < gravimeter >
             < lunar surface gravimeter >

For instance, compare Mexico City with its low latitude, high elevation and a smaller value of gravity to that of Helsinki, Finland. Helsinki is at higher latitude and lower altitude than Mexico City with greater value of gravity. A long-lasting world record in long jump (broad jump) was made at the Mexico City Olym-

221

pics. If it were possible to reproduce exactly the same long jump (broad jump) in Helsinki, the length of that same jump in Helsinki would have been an inch or so (a few centimeters) shorter. **Should the records be adjusted/computed for a standard gravity value?**

Google: < **athletic world records** >

< **markun kotisivut** > This is the home page (markku's homepages) of a Finnish gentleman keeping athletics records. Text is also in English. Click on the underlined event getting its statistics. For instance, there is 'Javelin' and 'Javelin, the new model'. To convert metric distances to English units, use 1 meter = 100 centimeters = 39.37 inches = 3.28083 feet.

Different gravity values at the spot (stadium) **of the event certainly have a significant effect on athletic performances and on the world/ Olympic/national records.** For instance, a weightlifter can lift more at the high topographic altitude of Mexico City than in Ottawa, Canada, Stockholm, Sweden, Helsinki, Finland, or at many other places with greater values of gravity.

To fairly take the gravity effects into account, World and Olympic Records could easily be corrected for any gravity value in any athletic event including **weight lifting, javelin throw, broad jump, high jump, shot put, hammer throw, pole jump and so on.**

The following considerations and computations of athletic records were published by **Weikko A. Heiskanen in the** *Scientific American*, **September, 1955, Vol. 193, No. 3,** in a seven-page article starting on page 164. For more detailed information, one may have to go a library because **the mentioned article could not be found on the Internet.**

Some examples based on Heiskanen's article follow.

41.1. Javelin Throw

On May 25, 1996, Ján Zelezny, CZE, threw his javelin 98.48 meters (323.10 feet) in Jena, Germany.

Google: < jan zelezny >

The nearest approximate latitude in Heiskanen's table on page 170 for Jena is 50.3 N (Berlin). When compared to an imaginary, exactly the same javelin throw at 40 N latitude and at the same topographical altitude in Columbus, Ohio or in Southern Italy, where the gravity is less than in Jena, Germany, the correction factor is 0.001119, resulting in an 11-centimeter (4.33 inches) difference.

This means that the same javelin throw, at 40-degree latitude at the same topographical elevation would have been longer by 11 centimeters (4.33 inches), or the throw would have been 98.59 meters (323.46 feet). A greater topographic altitude in Columbus, Ohio than in Jena, Germany would further make the correction even greater than the 11 centimeters (4.33 inches).

Recall that the gravity should be measured at the place where the javelin flies. Also, one might want to consider the air pressures also as well as accurate measurements of applicable wind velocities.

If Jan Zelezny would have made the same javelin throw in Helsinki, Finland, where the gravity is stronger than in Jena, Germany, the correction factor (compared to 40 N latitude) is 0.001760, resulting in a correction of -17 centimeters (-6.83 inches). The length of the same javelin throw in Helsinki would have been 'only' 98.31 meters (322.54 feet).

So, what should Jan Zelesky's world record javelin throw be on May 25, 1996?

In Jena, Germany he threw 98.48 meters (323.10 feet).

At a 40 N latitude in Columbus, Ohio the same throw would have been 98.59 meters (323.46 feet).

At a 60 N latitude in Helsinki, Finland the same throw would have been 98.31 meters (322.54 feet).

Actual gravity measurements at the stadiums where the javelin was thrown could modify these numbers by small amounts.

Centimeters and inches do count for world records in the Javelin throw.
There is something here for the Olympic Committees to consider.

Similar differences caused by gravity variations could be considered in other athletic events where world and Olympic records are kept, such as pole vault, hammer throw, shot put, high jump, long jump, weight lifting and in some other competitions where the intensity of gravity affects the results.

Google: < olympic records >

42. GRAVITY MEASUREMENTS

Isaac Newton (1643-1727) discovered the law of universal gravitation. The law of gravitation has been in continuous operation long before humans appeared on Earth. **It is really God's/Physics'/Nature's Law of Universal Gravitation.** However, Isaac Newton deserves the honor of being the first person wise/clever enough to find the existence of that law.

Universal Newtonian gravitation is the mutual pull (attraction) between two masses. It is an acceleration (= increase in speed) these two bodies (masses) try to start getting together (moving, falling) for a collision.

Earth's gravitational pull is called Earth's gravitation.

Earth's gravity is the vector sum of its gravitational pull and centrifugal acceleration. Centrifugal acceleration is generated by Earth's daily rotation (spinning). Centrifugal acceleration reduces the gravitation by small amounts.

Earth's average gravity = one G = 9.81 meters per second squared (m/s/s) = 32.2 feet per second squared (ft/s/s) is slightly reduced by Earth's centrifugal acceleration due to Earth's daily spinning once around its N–S spin axis in 24 hours. This centrifugal acceleration 'cancels' a small part of Earth's gravitational pull. Maximum centrifugal acceleration is at Earth's equator with a value 0.0034 meters per second squared (m/s/s) = 0.011 feet per second squared (ft/s/s), or approximately 0.03 % of the Newtonian gravitational attraction (pull).

If Earth would spin once around in 1.4 hours (instead of 24 hours), the centrifugal acceleration would totally 'cancel' the gravitational attraction at the equator making all masses weightless there. Of course, there would be numerous other consequences such a more flattened Earth with much deeper ocean waters near the equator and shallower waters at high latitudes.

225

Google: < isaac newton >
                  < centrifugal acceleration on rotating earth >

Earth's gravity is always with us. It attracts us constantly and relentlessly to the ground giving us our weight (lbs, kg) and keeps us from floating away to the oblivion of the surrounding vacuum of space.

Gravity pulls us constantly, as well as all other material masses/objects, toward Earth's center, whether we are in America, Australia or anywhere else.

Gravity keeps Earth's atmosphere pressing/hugging us with an average pressure of 1.03 kilogram force per every square centimeter of our skin = 14.7 pound force per every square inch of our skin. We like it, don't we!

Gravity pulls the atmospheric gas molecules toward the ground/waters pressing Earth's surface down with its weight on every square inch (centimeter) of Earth's surface.

One may think that our feet/bodies are pressing the ground/floor/trap-door/some support under us. A more correct way of thinking/saying would be to realize that **it is the gravity of Earth that pulls/attracts our body mass toward the center of Earth causing a temporary pressure increase against the ground under our bodies.**

No person (on the ground) can increase her/his pressure against the ground over a period of short time without either carrying/supporting some extra mass or without trying to push/pull some ground supported object (ladder, house floor from a crawl space, pull a fence post up, etc.). Even then, only a part or the 'whole thing' of that object's pressure against the ground is transferred from the object's earlier support points to underneath the 'struggling' person.

Earth's gravity keeps Earth satellites and 'our' Moon in their orbits.

Google: < the rotating earth >
                  < standard atmospheric pressure >

Even the Shuttle/ISS astronauts orbiting Earth are under the pull of Earth's gravitation *toward the center of the Earth*. **Earth's gravitation keeps their satellite in orbit** *compensating* **for** *their centrifugal acceleration radially away from Earth.* The centrifugal acceleration comes from flying in approximately 6,800-kilometer (4,200 mile) radius circle around Earth at a speed of approximately 7 kilometers (4.3 miles) per second. Mattresses are not needed for astronauts to sleep on.

Without Earth's gravitation, they would fly into oblivion. Without centrifugal acceleration, they would fall down to Earth.

If there were no pull of gravity here on Earth, a slightest jump/push would send us into outer space. Of all the atomic forces, **gravitation is considered to be the weakest force;** at least it is much weaker than the electric and magnetic forces. Gravitation everywhere on Earth is just strong enough to keep us and all other earthly masses, including the atmospheric gases, comfortably close to the ground with a reasonable force. Atmospheric gases get into every open cave, nook and cranny. We have breathing air everywhere near Earth's surface, although it may get a little thin on top of Mount Everest! The Moon's gravitation is too weak to keep an atmosphere, and the gravitation of planet Mars is not much better.

Google: < martian atmosphere >
             < types of particles and forces >

Some neutron stars and 'black holes' have gravitational fields/forces millions of times stronger than the gravity on Earth. The gravity on the Sun's surface is a bone-crushing 27 G. Some neutron stars spin at a rate of 30 rotations per second having tremendous centrifugal accelerations at their equators.

One may stand at the same spot where a wheel of a car was a moment earlier. **Gravity at that spot is essentially the same** whether you stand on the spot or if a car wheel is parked on that spot. The car wheel presses the ground with a greater force than your feet because the mass pushing that wheel down is greater than your mass.

As has been mentioned, the average numerical value of acceleration of gravity on Earth's surface is 9.81 meters per second squared (m/s/s), or 32.2 feet per second squared (ft/s/s), or one G, or 981 Gal, or 981,000 milligals, or 981,000,000 microgals. The unit Gal honors Galileo Galilei (1564-1642).

Google: < gravity meters >
             < gwr instruments >
             < neutron stars >

The direction of gravitational attraction/gravity/plumb-bob is toward Earth's center of gravity only with small angular variations (deflections), which usually are less than one minute of arc. One minute of arc is an angle in which 6 centimeters (2.36 inches) is seen at 206 meter (676 feet) distance.

The direction of the centrifugal acceleration is in the plane of the parallel circle of latitude or perpendicular to Earth's spin axis.

To understand gravitational pull one must deal with acceleration. One can physically feel changes in acceleration, when in an airplane accelerating/braking on a runway, in a car when increasing/decreasing its speed or riding a merry-go-round. Fast-flying fighter pilots and the airplane frames/wings making tight turns cannot endure accelerations much greater than 10 G for a very long time. That kind of acceleration puts a strain/bending/twisting forces on the fuselage and wings of the airplane in addition to the breath-holding strain on the pilot affecting breathing and other 'things' in the human body.

Acceleration means increasing the speed (pedal to the metal)**, and deceleration** (braking) **means decreasing the speed.** If the speed of a vehicle on a level road is constant, the acceleration of the vehicle is zero, i.e. the vehicle is not increasing its speed nor is it slowing down.

Some gravimeters (gravity measuring instruments) can measure the intensity of gravity to the nearest microgal, which is one part in a billion of the total Earth's gravity. To understand what this means, consider a gravimeter placed on a floor and then on a one-foot high stand (0.3 meters) directly above. **The measured gravity will be approximately one tenth of a milligal (100 microgals) less on the one foot high stand because the distance of the gravimeter to the center of Earth is one foot greater than it was from the floor**. This is simply in accordance with Newton's Law of gravitational attraction.

If the height of the gravimeter stand had been only three millimeters (0.12 inches), the measured gravity would have been one microgal smaller on the stand than the measured value on the floor. There are gravimeters which are capable of this kind of precision, and there are scientific reasons why this kind of precision is necessary.

Google: **< physics: newton's law of gravitation >**
            < acceleration in physics >

< centrifugal force >
< earth's gravity >
< acceleration >

Using a high school student's pocket calculator, one can easily compute the value of gravitation at a distance of 6,371,200 meters (average Earth's radius) and also at one-meter greater elevation of 6,371,201 meters from the Earth's center. These two values will differ by 0.3 milligals per that one-meter distance difference, or 0.1 milligals per one-foot elevation difference, as stated. The numerical calculation goes as follows:

In the formula of Newton's Law of Gravitation, in the numerator, use the **product of the universal gravitational constant times Earth's mass.** That product is **3.985115 x 10E20** when using units of centimeters, grams and seconds. Divide this product first by 6,371,200,000 centimeters squared and then also by 6,371,201,000 centimeters squared. Find the difference of the two numbers as stated.

The necessary numerical values are available at the following Internet sites.

Google: < nist >
                 < the nine planets >
                 < newton's law of gravitation >

In the first case, one will get Earth's gravitational attraction at a distance of 637,120,000 centimeters (6,371.200 kilometers) from Earth's center as **981.7444329 gals. In the second case**, one will get Earth's gravitational attraction at a distance of 6,371,201,000 centimeters (6,371.201 kilometers) from Earth's center as **981.74412476 gals.** The difference of these two values is 0.0003082 gals or 0.3 milligals. **This simple calculation shows that near Earth's surface, gravity is decreasing by approximately 0.3 mgals per one-meter increase in elevation.** The number 0.3 milligals per meter is called the **gradient of gravity** at sea level elevation or also the 'Free Air Correction'. Its numerical value gets smaller at higher altitudes.

According to Newton's Law, gravitation decreases with increasing distance (square the distance) from Earth's center. **Centrifugal acceleration also causes the gravity to decrease** on Earth's surface ever so slightly at higher elevations, i.e. increased distance from Earth's spin axis increases the centrifugal acceleration.

Once more, the **average value of acceleration of gravity on Earth is 9.81 meters per second squared, or 9.81 Gals, or 1 G, (32.2 feet per second squared)**. Centrifugal acceleration is zero at Earth's poles and largest at the Equator, where its value is approximately 0.3% of the total gravitational pull. **Centrifugal acceleration reduces gravity at the Equator to 9.78 Gals. At the poles the value of gravity is approximately 9.83 Gals.**

Another main reason why gravity at the poles is slightly greater than at the Equator is the fact that the poles are a little closer to the heavy central core of Earth because the polar radius is a little shorter than the equatorial radius. These details will not be discussed in this book any further.

43. GRAVITATIONAL PULLS ON THE MOON

The average distance between the centers of the Earth and the Moon is approximately sixty times Earth's radius or 1.2 light-seconds. The intensity of the pull of **Earth's gravitation keeping the Moon in its orbit around Earth is only one part in 3,600 (3,600 = 60 x 60) of one G, or about 270 milligals or about 0.000270 of one G.**

Google: < newtow's law of gravitation >
 < moon fact sheet >
 < kepler's laws with animation >
 < ernest w brown >

Again, the value of Moon's gravitation (gravity) on its surface is needed. It is approximately one sixth of Earth's gravity at sea level. Compared to Earth, the Moon's centrifugal acceleration is very weak because the Moon spins around its spin axis approximately only once a month and its radius is only 27% of Earth's radius. Therefore, the **Moon's gravity is almost entirely pure gravitation.**

The Sun's gravitational pull (attraction), keeping this planet Earth and the Moon in their annual solar orbit** here at one astronomical unit = 1 AU = 149,598, 000 km = 93,000,000 miles distance from the Sun, is approximately **590 milligals or about 0.0006 of one G. Note that Earth's centrifugal acceleration when flying at about 30 kilometers (18.6 miles) per second in the Sun's gravitational field at 1 AU distance from the Sun is also the same 590 milligals.** Centrifugal acceleration equals the gravitational attraction keeping the radial forces (moving 'this and that way') in balance.

Adding up, as the Earth is tugging the Moon at 270 milligals, at the same time the Sun is tugging both the Earth and Moon at approximately 590 milligals. Ernest W. Brown (1866 – 1938) is one astronomer who spent years in computing his accurate lunar tables. Present-day astronomers can do the same and more with their computers in a 'jiffy'. How lucky we are!

43.1. More on Physical Meanings of Accelerations

As has been mentioned, gravitation and gravity are accelerations. When a body (mass) is in a gravitational/gravity field, the field will exert a force upon the body (mass), which is proportional to the mass of the body. For instance, a baby is pulled down by Earth's gravity by a weaker force than an adult since a baby's mass (weight) is less than an adult's (weight). **Weight is mass times acceleration.**

Let's go upward 'playing astronauts' and consider only 'ball park' size numbers. If one takes off in an imaginary Shuttle Orbiter with a constant acceleration of 1 G, it takes 714 seconds (11 minutes 54 seconds) to reach the orbital velocity of 7 kilometers per second (4.3 miles per second) when the velocity is increasing by 9.81 meters (32.2 feet) per every second after launch. If the constant acceleration would be 2 G, the mentioned orbital velocity would be reached in half the time, or in 357 seconds or in 5 minutes 57 seconds. In addition of getting the Shuttle up to its speed, it must also be lifted to its orbital altitude.

A freely falling body near the Moon's surface is increasing its falling velocity by approximately 1.6 meters per every second (5.4 feet per second) until it hits the Moon's ground or something else. After ten seconds in free fall on/near the Moon, the speed has increased to 16 meters per second (54 feet per second) or to 36 miles (58 kilometers) per hour.

Google: < free falling objects >

Because this Earth is not a homogeneously uniform sphere, the gravity varies slightly at sea level (zero elevations) over and under the 1 G value.

At higher topographical elevations gravity is also less than at sea levels.

At higher latitudes gravity is also greater than at lower latitudes.

These gravity variations from the normal gravity are called gravity anomalies. The actual value of gravity depends on several 'things'. It depends on the location, i.e. on geodetic (geographical) latitude and longitude, topographical elevation, Earth's ellipsoidal shape, the mass irregularities on and within the body of Earth and also slightly on the attractions of the Moon and Sun (tidal effects).

Google: < international gravity formula >

43.2. GRAVITY ANOMALIES

In addition to the geometrical features (size and shape), Geodesy (Geodetic Science) also studies **Earth's gravity field, which varies with the location and elevation.** The value of acceleration of gravity on Earth is mostly between 9.77 and 9.83 meters (32.1 and 32.2 feet) per second squared. **The smallest gravity values are generally at high elevations near the Equator, and the largest values are generally at low elevations at high latitudes.**

These variations in gravity anomalies caused by irregular mass distributions inside the Earth may seem small, but they are measurable and **important in many applications. For instance, these gravity anomalies cause perturbations in the orbits of artificial satellites, which are taken into account in satellite orbit determinations/maintenance.**

One part in accurate computations of satellite orbits must consider the variations in gravity (gravity anomalies) all over this Earth. **Gravity anomalies are also used in prospecting for minerals, ores, oil and other materials buried in the ground.**

Google: < gravity anomalies >
             < prospecting using gravity anomalies >

It is the gravity that keeps all of us 'clinging' to Earth whether we are walking (on a reasonably horizontal non-slippery surface) in America, on a seemingly vertical surface around the Eastern Mediterranean areas or are seemingly walking almost upside down in Australia.

The distance between the US East and West coasts is approximately 3,000 miles (5000 kilometers, or three 15-degree time zones) meaning that the vertical directions between US East and West coasts differ by about 45 degrees, which is a steep tilt.

Gravity is also important for our general health. For instance, standing upside-down (like in yoga) for long periods of time can have serious health consequences. Astronauts returning from long periods in weightlessness have difficulties even in walking on Earth right after landing.

God/Creator made the human body fluids to be balanced under normal (1 G) Earth's gravity, whether we are in Australia or in America. Humans (especially fighter pilots) are sometimes tested in centrifuges for tolerances of greater accelerations than 1 G. Even breathing becomes difficult, and many faint when the acceleration approaches 9 G.

Earth's gravity is an acceleration producing an attracting force trying to pull us and all other masses against/toward/down to the ground. If there is nothing to support us – such as a trap door – we will go down so that our speed is increasing by approximately 10 meters (33 feet) per every second of the fall until we land on something.

44. Geoid (= Mean Sea Level) Undulations

Geoid undulations are caused by gravitational attractions of non-homogeneous ground level and underground masses. They are rather permanent humps and depressions in the mean (= average) sea level. Their size is up to plus-minus 150 meters = 500 feet. A large swimming pool can be an example of this effect. If one cubic meter (1.3 cubic yards) of lead with a mass (weight) of 11.35 metric tons is placed at the center of the pool bottom, and one cubic meter of water is removed from the pool, then the original water surface will be distorted so that the water will be deeper at the center of the pool than at its sides and corners. The volume of the lead with its 11.35 metric ton mass replaced the same volume of water with a mass of one metric ton. Therefore, the volume of the lead has a mass 10.35 metric tons greater than the original volume of water, and it pulls all water molecules in the pool 10.35 times stronger than the original one cubic meter of water, gathering a hump of water over the middle of the pool.

Due to underground mass irregularities mean sea level (=Geoid) of all Earth's oceans does not have a perfect mathematical rotational roundness. It undulates over and under by up to 150 meters (500 feet) of a well-chosen mathematical rotation Earth Ellipsoid (of revolution) with perfect roundness. These undulations can be more abrupt (steeper) across deep oceanic trenches such as the Mariana Trench in the Western Pacific or in land areas over mountain ranges with rough topography, for instance, the Grand Canyon. Usually Geoid undulations cover large areas such as the center of the North Atlantic Ocean with some local (permanent) ripples on the undulations. For instance, a ship traveling along Earth's Equator (or along some other parallel of latitude) may sometimes be a couple of hundred meters (650 feet) nearer to Earth's center than at some other places.

Earth Ellipsoids are sometimes called Spheroids because the Earth Ellipsoid is almost a sphere. Its equatorial radius is approximately 6378 kilometers (3963 miles), and its polar radius is approximately 6357 kilometers (3950 miles) or only 0.03 % shorter than the equatorial radius. The Geoid approximates this mathematical surface.

Google: < spheroids. ellipsoids and geoids >
             < geoid and equipotential surfaces >
             < the geoid >
             < topex >
             < geoid map >

44.1. PLUMB LINE DEFLECTIONS AFFECT ASTRONOMICAL COORDINATES

Due to Earth's mass/density irregularities (flat lands, mountains, valleys, lakes, ore deposits, etc.) inside and on Earth's surface, **plumb lines are usually deflected** by small amounts in any compass direction (azimuth), usually not much more than one minute of arc an angle formed by 6 centimeters (2.4 inches) at a distance of 206 meters (676 feet). One minute difference in astronomical latitude corresponds to one nautical mile.

Google: < deflections of the vertical >
             < angular degrees minutes and seconds >
             < nautical mile >

Plumb line deflections are called Deflections of the Vertical. Plumb lines are perpendicular to the bumpy average sea level, the Geoid. Latitude/longitude determinations by astronomical methods rely on the plumb line at the location of the measuring instrument. Therefore, astronomical latitudes differ from the geodetic (geographical) latitudes by the North-South component of the deflection of the vertical. The astronomical longitudes differ from the geodetic (geographical) longitudes by the East-West component of the deflection of the vertical.

Astronomical positioning of latitudes/longitudes may differ by a mile (one nautical mile is 60 seconds of arc) from the geodetic (geographic) and GPS latitudes/longitudes. **At the present time, continental distances can be accurately measured to the nearest few centimeters/inches depending on the methods used**.

Astronomical positioning methods for geographical latitudes and longitudes were generally the best methods in use before 1950. Long distance global positioning was not very accurate before 1950. The coordinates for distant islands and continental separations often differed by one mile or more from the present day much more accurate geodetic (geographical) coordinates.

Some 50 years ago astronomical-positioning instruments were precise enough to determine (the astronomical) positions rather easily to the nearest correct second of arc, which on the ground corresponds to approximately 100 feet or 30 meters. **The same instruments were also precise enough to determine relative coordinates on land areas down to millimeter accuracies. The problem was that the deflections of the vertical were not well known. Astronomical positioning instruments cannot be set up** (leveled) **on the oceans.**

One arc-minute distance on the Earth's surface is called one nautical mile. One arc-degree arc-distance on the Earth's surface is 60 nautical miles = 69 US statute miles = 111.1 kilometers. If the end points of one nautical mile arc on the Earth's surface are 'connected' to the center of the Earth, the angle between these two radii is one minute of arc. For comparison, as state above, the end points of sixty millimeters (2.36 inches) at a distance of 206 meters (676 feet) also extend an angle of one minute of arc.

For another comparison, as seen from Earth, **the apparent diameter of the Moon as well as the Sun extends an angle of approximately 0.5 degrees** or 30 minutes of arc. When the Sun or/and Moon are near the horizon, they look bigger than when they are higher in the sky, which is only an illusion of the human eye. The measured horizontal diameters of the Sun and Moon are at all altitudes approximately 0.5 degrees = 30 minutes of arc large. – Refraction comes into play if vertical diameters are measured.

Google: < solar semi-diameter >

     < nasa eclipse home page >

44.2. ELEVATIONS AND DEPTHS ARE MEASURED FROM THE GEOID

The **reference surface for elevations (up) and depths (down) is the mean sea level = Geoid.** It is a level (= equipotential) surface of zero topographical elevations. The mean (average) sea level is a level (horizontal) surface without having any uphill/downhill slopes. Topographic elevations are measured up/down along a plumb line from the Geoid including its continuation under the continents/islands. A common name for topographic and bathymetric numbers (feet/meters) is **Vertical Datum.** GPS, various types of spirit, automatic and trigonometric leveling instruments are used for leveling operations. GPS, some satellites and tidal gages are used to measure the instantaneous sea levels.

Google: < 1-mean sea level, gps, and the geoid >

     < vertical datum >

     < topex >

     < leveling instruments >

     < trigonometric leveling >

http://www.astrologyclub.org/articles/nodes/nodes.htm

The mean sea level has daily, monthly and annual variations. In addition to the daily tides, the gravitational pulls (tugging) by the Moon and Sun of all earthly masses (ocean floors/waters included) have also a major periodic effect on the mean sea level. The main **cycle of the lunar effects (tides) is 18.6 years long**; therefore, **mean sea levels must be observed over a multiple of 18.6-year periods to obtain good average values for the mean sea level**. The periodic solar gravitational effects average out in one year. There are over a dozen reasons why the sea level is not constant anywhere.

Google: < sea level variations >

The long-term rise in the mean sea level during the last 50 years has been approximately **2 millimeters per year.** As was mentioned, there are numerous reasons for the global sea-level rise. Speculations by the uninformed general news media seem to overestimate/favor the tiny amounts of melting of some glaciers here and there around the globe. The problem with that type of speculation is the fact that 99% of Earth's land-supported ice masses are in Antarctica and Greenland, and nobody knows yet whether those large ice masses (ice budgets) are increasing or decreasing. Some glaciers are melting, and some are accumulating more ice masses.

Melting of floating sea ice – such as the ice sheet and icebergs in the Arctic Ocean – **has no effect on the sea level by itself.**

The seldom-considered thermal volume change of ice/water is another matter. It will only be mentioned here. A floating iceberg to be melted has a lower salt concentration in it than the surrounding seawater. According to the Archimedes' Principle, **the floating iceberg displaces a volume of salty seawater with the same mass as the iceberg has.** When an iceberg turns to water, its melted water has a smaller volume but the same mass as the original iceberg had. As the iceberg melts, the temperature of its melted water is 0 Celsius = 32 Fahrenheit = 273.15 K. Later on, this new melt mixes with all surface waters in all oceans having an average temperature of approximately 17 C = 62.6 F. When the new melt warms up to 17 C = 62.6 F, its volume increases (sea level rises). The average temperature of deep ocean waters constituting 90% of all ocean waters is between 0 C to 3 C = 32 F to 37.5 F.

Google: **< average temperature of ocean waters >**

Most of the icebergs in North Atlantic and the Arctic Ocean have calved from Greenland. **The total *water equivalent* snow accumulation on the total Greenland's ice sheet is** approximately **30 centimeters (one foot) per year (2004).** At some places, the depth of the accumulated snow may be up to ten times that much. That amount, 0.3 metric tons on every square meter of Greenland, is greater by far than many mountain glaciers combined under the study of the general news media. At the present time satellites are measuring the global ice budgets of land-supported ice sheets. Hopefully in the near future, sound scientific answers about the Antarctic and Greenland ice budgets will become available.

It is obvious that if someone knows what happens to the 1% of the land-supported ice masses but does not consider the remaining 99%, it is not wise to draw general global conclusions.

Google: < mean sea level >
             < geoid >
             <ngs tidal gages >
             <usgs tidal gages >
             <vertical datum >
             < ngs faqs >
             < usgs leveling >
             < ngs leveling >
             < noaa/ceob tide glossary >
             < mean sea level, gps and the geoid >
             < mean sea level variations >
             < highest tides >
             < ice and snow sublimation >
             < 10(ae) glacial processes >

Topographic maps have elevation contour lines of the ground. Bathymetric maps have water depth contour lines. Contour lines are some multiples of ten, hundred or thousand meters/feet/fathoms, in addition to some point elevation/depth numbers here and there. For instance, Denver, Colorado is the 'one mile-high city', and the Titanic rests on the ocean floor at a depth of approximately 12,000 feet = 2,000 fathoms = 3,700 meters.

Google: < usgs mapping information topographic map symbols >
             < ngdc-bathymetry, topography and relief >

The mean sea level is 'the water level' for the whole Earth. To get topographic elevations on the Moon or on planet Mars is another problem due to the absence of liquid water/any other liquid.

By definition, **the plumb lines (= verticals) are perpendicular lines to level/horizontal/equipotential surfaces** everywhere around the Earth. **Geoid is only one of many equipotential (level) surfaces**. Every stair step has its own equipotential (level) surface. Water runs in rapids and in lazy rivers because the plumb lines are not perpendicular to the sloping water surface. **Water at rest** (or another liquid) in a glass, puddle, pond, lake or an ocean is a part of its own equipotential surface. **Equipotential (level) surfaces are almost spherical mathematical shells surrounding the whole Earth.** Equipotential surfaces exist also inside Earth.

This book will not go into the details of Physical Geodesy/equipotential surfaces. Some Internet sites are given for those who desire more detailed information.

Google: < physical geodesy is everywhere >
            < features of sea level change >
            < nasa-as sea level rises, beaches shrink >

Accelerations of gravities/gravitations acting upon all masses in their attraction-fields produce forces that have made all stars, planets and larger moons almost spherical in shape. The Sun, planets and moons have their own gravitational attractions and their own equipotential (level) surfaces. Acceleration of Earth's gravity together with the gravitational attractions of the Moon and Sun make the Earth's instantaneous ocean surface to be the instantaneous almost spherical-curved sea-level surface with small humps and depressions.

In addition to the tides, winds, atmospheric pressures, inertia of water flow and irregular ocean floor topography affecting water flow, Coriolis forces, water temperature variations, occasional tsunamis and a dozen other reasons disturb the instantaneous sea level. The **Geoid is the mean (average) sea level.**

Google: < usgs open file report 96-000 >
            < coriolis force >
            < sea level changes >

There are many reasons why the instantaneous water depths in the oceans change constantly. For instance, in areas of high atmospheric pressure over the ocean surface, the ocean water flows to surrounding areas that are under a lower atmospheric pressure. Hurricanes provide a good example of this effect. In the eye of a hurricane the air pressure is low, and the ocean water depth is usually deeper by several meters (yards) in and near the hurricane's eye than in the surrounding areas. This effect is somewhat similar to sucking a part of the air out of a drinking straw dipped into a glass of water reducing the air pressure inside the straw. The higher pressure on the water surface around the straw pushes the water up in the straw.

45. Earth's Detailed Size and Shape

The Flat Earth Society does not have very many followers anymore, although in everyday life for most of us, our lives revolve mostly within the horizon around our homes. Seldom do we consider (astronauts and some airplane pilots do) the 'downward' curvature of Earth in all compass directions over longer distances from our homes, although we know that this **planet Earth is round.**

Some textbooks describe the method used by **Erathostenes** (275 BC – 195 BC) measuring **the length of Earth's circumference.** It is believed that his result was only 16% off from the modern value.

Google: < erathostenes' method >
           < noaa the elements of geodesy >

Isaac Newton (1643 – 1727) correctly **suggested that this Earth is a flattened 'sphere' at the poles.** Newton knew that Earth rotated, and he understood the effects of centrifugal acceleration on the body of Earth

When Earth is considered to be a sphere, its average radius can be taken to be 6371 kilometers = 3959 miles. **When Earth is considered to be a well-fitting ellipsoid of revolution**, its equatorial radius is approximately 6378 kilometers = 3963 miles, and its polar radius is approximately 6357 kilometers = 3950 miles.

The Earth Ellipsoid is a mathematically chosen rotation ellipsoid surface approximating Earth's mean (average) sea level (= Geoid and its continuation under the islands and continents) for the entire Earth. The spherical Earth with its average radius of 6371 kilometers = 3959 miles has also almost the same volume as a well fitting reference rotation Ellipsoid, sometimes called less precisely Earth ellipsoid of revolution. Recall that **Earth rotates around its N-S spin axis, and it revolves around 'our' Sun.**

Google: < earth: geodetic and geophysical data >
                 < the earth as an ellipsoid >
                 < us geological survey >

Topographic maps give the elevations, latitudes and longitudes of ground-level points above and under the mean sea level = MSL = Geoid.

Bathymetric maps give water depths, latitudes and longitudes for bodies of water.

Google: < national geophysical data center >
                 < bathymetric maps >

There are also maps that give the **sea level topography or the undulations of the Geoid = undulations of the mean sea level above and under a chosen Earth Reference Ellipsoid. Geoid undulations are usually not greater than 150 meters (500 feet), which is only up to 0.002 % of the equatorial radius of Earth**. At most places on Earth these Geoid undulations are smaller than plus-minus 30 meters (100 feet).

Google: < sea level topography >
                 < topex global sea level >
                 < geoid undulations >
                 < earth ellipsoid measurements >

The small Geoid undulations show how very smooth this Earth is over its ocean areas covering 71% of Earth's surface, when comparing the plus/minus 150 meters to 6,371,000 meters. The highest mountaintops and the deepest ocean trenches show the maximum deviations of the ground and the sea floor from Earth's smooth roundness. The numerical values are repeated in the following:

The **deepest ocean water is in the Mariana Trench** in the western Pacific Ocean. That water depth is 35,813 feet (10,915 meters) according to one measurement. This lowest dip in the sea floor is only 0.17 percent of Earth's mean radius.

Google: < deepest ocean >
                 < exploring deep ocean floor >

The highest mountain peak is Mount Everest at 29,035 feet (8,850 meters). This highest ground elevation is only 0.14 percent of Earth's mean radius.

Google: < highest mountain peaks of the world >
             < highest mountain peaks >

How perfectly spherical is this Earth? On a tabletop globe with a diameter of 10 inches (25 centimeters), Mount Everest would show up as a hump of 0.17 millimeters (0.007 inches) high, a relatively small hump, considering the whole globe. The Mariana trench would be a 0.22 mm = 0.009 inch deep dent.

45.1. Earth's Shape in the Room Size Balloon Model

Only a few people notice in their daily lives the fact that this planet Earth is just a small, round, spinning globe, continuously flying through the vacuum of space on its annual roundtrips (orbits) around 'our' Sun at an average orbital speed of 30 kilometers (18.6 miles) per second. Because there are 3600 seconds in every hour, by simple multiplication the same speed can also be expressed as 107,200 kilometers per hour = 66,600 miles per hour.

It is the centrifugal acceleration resulting from Earth's daily (diurnal) rotation (spinning), that has caused the body of this Earth (ground and mean sea level) to bulge out at the Equator, i.e. the equatorial diameter is a little longer than the polar diameter. The equatorial radius is 21.384 kilometers = 13.288 miles longer than the polar radius.

Although this deviation from a perfect sphere is small, it is a very important feature in accurate positioning (latitudes and longitudes) and map making. This **deviation from a perfect sphere is called flattening of the meridian ellipse, also called the eccentricity of the meridian ellipse or simply the flattening of Earth.** (The Flat Earth Society went all the way to the pancake-shape!) All meridian ellipses on a rotation ellipsoid are of the same size and shape; therefore, they all have the same flattening and eccentricity. When any meridian ellipse is rotated around Earth's spin axis, that meridian ellipse traces the surface of a rotation ellipsoid.

Google: **< earth: geodetic and geophysical data >**
                 < the earth as an ellipsoid >

One can get a good visual view, appreciation and understanding of the amount of Earth's flattening, and some other terrestrial quantities/dimensions by **considering the previously described room-size Balloon Model of Earth depicting this globe in a scale of one to five million.**

Some organizations/libraries have such large globes of Earth on display. The scale (1 to 5,000,000) of the room-size Balloon Model is such that when any actual linear distance on the model globe is multiplied by five million, one gets the corresponding actual linear distance on the actual Earth. Wall maps of the United States in scale 1:5,000,000 are quite common. The distance on such a map between the Pacific and Atlantic shores across America is approximately one meter (39 inches). That distance on the ground is five million times longer or about 5,000,000 meters = 5,000 kilometers = 3000 miles.

Google: **< welcome to the usgs – us geological survey >**

The spherical, room-size balloon model's diameter is one five millionth of Earth's diameter = 2 x 6,371,000 meters divided by 5,000,000 = 2.548 meters = 254.8 centimeters = 8.361 feet. This spherical balloon model was placed earlier in a room with a ten-foot (3.048 meter) high ceiling. Vertically, this spherical balloon touches the floor, and its top is 1.64 feet (50 centimeters = 304.8 – 254.8) from the ceiling.

Next, consider that this balloon will be slightly flattened so that its vertical diameter is reduced by 8.54 millimeters (0.336 inches) to 253.986 centimeters. (These numbers were so chosen that the flattening of the balloon would be close to Earth's actual flattening). Assume that the equatorial diameter remains at 254.84 centimeters (8.3609 feet).

Looking at this slightly flattened 8.361-foot (2.548-meter) diameter balloon from a ten-foot distance, one can hardly tell by just looking that its horizontal diameter is 8.54 millimeters = 0.336 inches wider than it is tall. This balloon model has the same flattening as the actual Earth with three-significant-number accuracy; therefore, it is a good visual model for the actual Earth on a scale of one to five million (1:5,000,000). This example shows how very close Earth is to being a perfect sphere.

46. Traveling on the Round Earth

Recall that the length of a meridian arc of one degree is 111 kilometers (69 miles) long. Proportioning the meridian/great circle distances, one finds that traveling on the Earth's surface in any compass direction (azimuth) by 31 meters (101 feet), the plumb lines at the end points of this distance make an angle of approximately one second of arc with each other. Similarly, **when driving a car in a north or south direction at a speed of 69 MPH (111 kilometers per hour), the latitude of the car is changing by one second of arc in every second of time**. At that speed (69 MPH) on a straight road, one nautical mile is covered in one minute of time. It can be said that a car is driven one (statute) mile around a city block, but a nautical mile usually implies a more straight-line distance used in navigation.

In some respects, many of us tend to have a **'Flat Earth Society Mentality'** although everyone knows that this planet is a spinning, round globe orbiting 'our' Sun in the emptiness of space. In our 'little worlds' **we tend to forget that we do live on a rather small round planet.** The 'Flat Earth Society' people do not take into account that their horizontal plane tilts as they travel around the globe. For instance, traveling between the East and West coasts in America, the tilt is approximately 45 degrees.

Wherever we are on the Earth's surface, we see only that part of the planet that is within our horizon, which from a ground location extends only a few miles in all compass directions (azimuths) around us. At some locations, distant mountaintops may tower above our horizontal plane.

Forgetting momentarily the all important reality that this Earth is round, it is somewhat unusual to think that people and ships on the opposite side of Earth seem to be able to walk/sail 'upside down'. The real world is such that near our antipode (20,000 kilometers, or 12,430 miles or 180 degrees away) the floors are parallel to our floors/ceilings or vice versa. The water glasses and the lakes are seemingly upside down without spilling any water out of those containers.

It is the acceleration of gravity that pulls us, as well as all masses including water and the atmospheric air toward the ground. All masses 'want' to go down (to a lower equipotential surface such as lower stair steps) **unless something supports and stops them from going even further down. Gravity 'sucks' us toward the ground 'like a magnet' no matter where on Earth we are.** Think globally! Every spot on the ground has its own 'down' and 'up' direction.

This is also the basic reason why 'our' Sun, planets and bigger moons are getting more spherical all the time as rocks, etc. are rolling down, not up 'on their own'. On the gas-giant planets, such as Jupiter and Saturn, solid bodies would sink a long way out-of sight into oblivion.

Google: < antipode on earth >

The antipode is at a maximum distance of 20,000 kilometers (12,427 miles) for any point along Earth's surface. Along Earth's surface one cannot go any farther away than that. If one continues beyond the antipode, one gets closer to the initial point from the other side of Earth. For example, the antipode of Corpus Christi, Texas is in the Southern Indian Ocean approximately halfway between Australia and Madagascar. The antipode of the North Pole is the South Pole. It is still 20,000 kilometers (12,427 miles) no matter along which meridian one travels from the North Pole to the South Pole. From Corpus Christi, it is still very close to 20,000 kilometers (12,427 miles) to its antipode, no matter in which azimuth one starts to travel from Corpus Christi while staying on a constant great circle. The ellipsoidal shape of Earth was ignored in these examples because the difference is 'small enough' for this type of an example, where millimeter accuracies are not needed.

This book does not cover ellipsoidal distances, geodetic lines = geodesics, which can be slightly shorter than the great circle distances. **A plane through Earth's center 'cuts' a great circle along Earth's surface.** The ellipsoidal meridian distance between the North and South Poles is also a great circle distance, and it is also a geodetic line and the minimum possible distance between the poles along Earth's surface.

Great circles on Earth's surface are 'cut' by planes going through Earth's center. Planes that don't go through Earth's center 'cut' (slice) the Earth's surface along small circles. Latitudinal parallel circles are examples of small circles .

Google: < great circle distances on earth >

For example, the distance from Corpus Christi, Texas to Eastern Mediterranean Sea is approximately 10,000 kilometers (6,200 miles). If a cruise ship from Corpus Christi has traveled to the Eastern Mediterranean, the view from Corpus Christi is that the ship is pointing in the 'straight down direction'.

As seen from Texas, the cruise ship in the Eastern Mediterranean looks like the last picture taken in 1912 of the sinking Titanic. From Texas' view, the Nile River seems to be flowing along a vertical wall from right to the left.

A good idea of a one-degree tilt or one-degree slope is visualized by the following example: If one end of a 57.3 ft long hallway is one foot higher than the other end, the floor is tilted by one degree. That hallway floor has a 1.7 % slope. A one-degree slope is one part in elevation difference divided by 57.3 parts of the distance, which fraction (1/57.3) prorates to 1.7 parts in 100 or 1.7%. **Also, if one end of a 57-inch long shelf on a wall is one inch higher than the other end, the shelf is tilted by one degree.**

Why was the length of the hallway chosen to be 57.3 feet? **The reason is that an angle of one radian is equal to 57.296 angular degrees** (of a full circle) **and also 206,265 angular seconds**. This makes it easy (?) to remember that a one-millimeter distance as seen from 206 meter (= 206,000-millimeter) distance extends an angle of one second. Also, one foot elevation difference in a 57.3-foot distance is a one-degree slope (angle), and that one meter elevation difference over 57.3 meter distance is also one-degree slope and that one-inch elevation difference over 57.3 inches is also one degree slope.

Google: < definition of angular unit of one radian >

For some slope comparisons, very few topographic slopes (down hills, up hills) on US Interstate Highways are steeper than 3%, or a three-foot elevation difference per one-hundred-foot distance. Whether the slope distance or the horizontal distance is used, makes only a small difference when dealing with small slopes. Another similar comparison is of a four-foot long shelf: if it is tilted by 0.5 degrees, one of its ends is 0.4 inches (one centimeter) higher than the other end, which tilt is usually noticeable to most people.

As is well known, the **length of Earth's circumference is 40,000 kilometers (24,855 miles). When claiming that someone travels around the world, the antipode of the starting point should be visited.** For instance, if one walks along a one hundred-foot (thirty meter) circle around the North Pole, crossing all the 360-degree meridians of longitude, **it would not** be correct to claim that a trip around the whole globe was made just by walking those 628 feet (192 meters) along a small circle around the North Pole.

It is quite possible that one car can be driven 24,855 miles (40,000 kilometers) completely inside some city's limits, but one would not call that traveling around the world. **As stated above, to truly travel around the globe, one should visit the antipode of the starting point, which is in the 'other' (southern or northern) hemisphere unless one travels along the Equator.** Otherwise, whoever claims to have traveled around the world by just cruising around 40,000 kilometers (24,855 miles) in a city, traveling only near some parallel of latitude, never crossing the Earth's equator, doing all the travel in the Northern or Southern hemisphere, is not telling a very convincing story.

47. TIDES

47.1. LUNAR AND SOLAR TIDAL EFFECTS ON ROTATING EARTH

Earth's tides originate from Earth's daily spinning (rotation) around its N-S spin axis in the gravitational pulling fields of the Moon and the Sun.

Earth's 'solid' ground, oceans and atmosphere are affected by the directional variations of the tidal pulls of all Earth's masses 'this way and that way', as the Moon and Sun are directly over various parts of this globe due to Earth's rotation and revolution.

The tidal effects of the Moon are approximately twice as large as the Sun's because the Sun is about 417 times further away from us than the Moon, although its mass is much greater. Recall that the distance to the Moon is about 60 Earth's radii long. The distance from the Moon to Earth's near side is 59 Earth's radii and the distance to Earth's far side is 61 Earth's radii, which difference is two parts in 60 = 0.0333. Similarly, the distance in kilometers from the Sun to Earth's near side is 149,600,000 – 6371 = 149,593,629 km and the distance to the far side is 149,600, 000 + 6371 = 149,606,371 km and their difference is = 0.000085. Once more, the tidal effects are caused by the variations in the gravitational pulls.

Google: < sun's mass >
             < mass of the moon >

The Newtonian gravitational attractions (gravity would be the wrong word here) of the Moon and Sun pulls every molecule on and in Earth (people included) *toward themselves* causing tides in the atmosphere, the oceans and also in the body of the more or less 'solid' Earth. Because Earth rotates 360 degrees in a day, during each day, this tidal pull can be up or down, east or west, or in some directions in between. This gentle 'jerking' of everything on Earth 'this way and that way' is an ongoing tidal process every day.

Google: < lunar tides >
< earth tides >
< solar tides >
< atmospheric tides >
< the moon and tides >
< tides in bay of fundy >

The total tidal acceleration (pull) on Earth varies between plus/minus 0.3 milligals, or plus/minus 0.0000003 times one G (G = 981,000 milligals = 32.2 ft/s/s). It may seem small but it causes 50-foot tides in the Bay of Fundy.

Google: < tidal corrections to gravity measurements >
< tidal effect on gravity >

Humans cannot feel these small tugs of attraction in their bodies, but nevertheless we are being pulled 'this way or that way' constantly. The tidal effects can be measured/observed in many ways.

As Earth spins 360 degrees per day, the direction from the center of the Earth to the center of the Moon changes by approximately 13 degrees per day, and the direction to the Sun changes by approximately one degree per day. The plane of Earth's orbit (ecliptic plane) is inclined by about 23.5 degrees to Earth's Equatorial plane, and the Moon's orbit around the Earth is inclined about 5 degrees from the ecliptic plane, and the Moon's and Earth's orbits are slightly elliptical. **For these reasons, the tides are not exactly the same on consecutive days.** It takes about 18.6 years for the Moon's and one year for the Sun's contributions to the tides to repeat themselves.

Take a look through Google at a tide-predicting machine designed by Lord Kelvin in 1873. It sure is nice to have modern computers!

Google: < moon's 18.6 years >
< noaa, tide predicting machines >

During our daytime hours we are nearer to the Sun than during the next night. The Sun's gravitational attraction pulls us 'up' during daytimes and harder against the ground at nights. The Moon does the same 'thing', only approximately twice as strongly.

At every instant of any day there is a meridian and a parallel of latitude, where the Moon is directly overhead between latitudes of 28.5 N and 28.5 S. Also, at every instant of any day there is a meridian and a parallel of latitude where the Sun is directly overhead between latitudes of 23.5 N and 23.5 S. Both the Moon and the Sun are trying to pull a hump of ocean water (and also the solid ground) toward their **sub-zenith points** on Earth, directly on the line connecting the center of the Earth to the center of the Moon, and the same goes for the Sun.

Google: < glossary of astronomical terms >

The points on Earth for which the Moon or Sun is directly overhead are called their **sub-zenith points.** At the time of a new moon, the Moon and Sun are over the same meridian, both pulling in the same direction. At other times of the month they are not in the same direction from Earth. If the Sun and Moon have the same declination at the time of a new moon, there will be a solar eclipse. **As Earth turns, the sub-zenith points will move 'around' the Earth.**

At the time of moonrise and sunrise at any point on Earth, the Moon and Sun try to pull the ocean waters (and everything on Earth) approximately eastward. At the times those celestial bodies set in the west, their pull is approximately westward. **These tidal pull variations knead the body of the solid Earth** constantly up/down/sideways including the ocean floors as well as all the upper layers (Crust, Mantle) of the body of the Earth. Ocean tides are most visible because the water in the ocean basins is relatively free to move and to slosh around by visible amounts. When the Crust/Mantle deform by a few centimeters, special instruments are needed to reveal those small movements.

The variable tidal accelerations (forces) slosh the ocean waters back and forth. **Irregular ocean shorelines and water depths interrupt the smooth continuous flow of the tidal waves around the globe in many ways**. In addition, the ocean waters are being dragged by the tidal forces over the more or less rugged ocean-floor topography with corresponding water-depth fluctuations, forcing the waters to flow 'this way and that way' creating swirls and gyres in all oceans. This all goes on 24/7/365.24/eons.

The same tidal forces (variations in gravitational accelerations, pulls) causing the ocean tides also **cause the so-called Earth Tides.** Earth Tides heave the continents and the sea floors (in addition to sea waters above the ocean floors) up and down every day **over a range of up to 40 centimeters (16 inches).** Is God/Creator rocking us constantly? Near the poles all tides are weaker (have a smaller amplitude) than nearer to the Equator.

As an example of the **tidal hydrostatic pressure changes on the ocean floors caused by the variable water depths,** consider the following example: Over one square kilometer of open ocean area (1,000,000 square meters = 0.39 square miles = 2,060 acres), the extra one meter (yard) thick high-tide water-layer presses (at high tides), or does not press (at low tides) the ocean-floor over that area by an extra weight of one million metric tons. At some shores, this ocean loading causes the shore to tilt by measurable amounts. This is sometimes called ocean-floor loading.

The Internet site of **The National Oceanic and Atmospheric Administration (NOAA) has several informative links explaining many aspects of tides**. It is a good site to read about the detailed explanations of the tide producing forces, and also why the tidal period is approximately twelve hours. The site explains why the tide producing force by the Moon is about twice as strong as the Sun's. The site gives some information on the trend of global sea-level variations.

Google: < noaa, our restless tides >
< mean sea level, gps, and the geoid >
< ocean tides >
< bay of fundy tides >
< earth tides >
< tsunamis >
< nodal points of ocean tides >
< lunar tides >
< solar tides >
< celestial coordinate system >

This book will not discuss the much **smaller tidal pulls coming from our 'nearby' solar planets Jupiter, Mars and Venus.** Earth is spinning in their gravitational fields also.

Open-ocean tides vary around the world. At low to mid-latitudes the height of open-ocean tides vary from one to three feet (30 to 100 centimeters) **with the usual frequency of two times per day.** In the Mediterranean, Baltic and Caribbean Seas daily tides have much lower amplitudes.

As has been mentioned, the difference of the gravitational pulls (accelerations) of the Moon and Sun at Earth's near and far sides is the tide-producing acceleration (force on the masses of Earth). The maximum variations in Earth's gravity caused by the Moon are plus/minus 0.1 mgal (milligals) and plus/minus 0.05 mgal by the Sun. Adding these two together, the maximum range of gravity variations is plus/minus 0.16 mgal, or with a total maximum range of approximately 0.3 mgal on the measured gravity at low latitudes. These tidal effects are smaller at high latitudes.

Tidal accelerations (forces, pulls) also 'wiggle' the plumb lines and tilt the instantaneous mean sea levels by small amounts. Maximum amounts of these plumb-line deflections are approximately 0.018 arc seconds caused by the Moon and about 0.008 arc seconds by the Sun. The corresponding maximum plumb-line deflections (tilts in the average sea level) are plus/minus 0.026 (= 0.18 + 0.06) arc seconds.

As an example, consider a hanging plumb bob at the time of moonrise and then do the same 12 hours later when the Moon sets in the west. The plumb bob is deflected eastwards at the moonrise, and westwards when the Moon sets. Instantaneous sea levels are also tilted by similar amounts as the plumb bob is deflected, because the plumb bob is perpendicular to the instantaneous sea level.

These tiny amounts in changes of gravity produce the mighty tides. Recall that 1 G (average gravity on Earth) is = 981,000 milligals.

In the open oceans the tides are not constant in all areas. The order of magnitude of open ocean tides is one meter (three to four feet) – more or less. Earth's Crust and Mantle are elastic and moveable. The Crust and the Mantle bend under the influence of the tidal forces of the Moon and Sun. As was mentioned, the range of solid earth tides can be up to approximately 40 centimeters (one foot) twice every day at some locations. **This means that we, together with the continents, are heaving up and down about that much twice every day.**

In addition to all that, the Crust and the Mantle heave up and down under the variable weight changes by the variable ocean water depths of deeper/ shallower waters above the ocean floors. This is sometimes called ocean-loading.

Ocean tides also produce strong water currents up to approximately 5-meters (15 feet) per second = 10 MPH = 16 km/h in many parts of the deep oceans, especially over and around seamounts, oceanic ridges and other rough sea-floor topography. Recall that the best 100-meter dash sprinters can run about10-meters per second.

Tides also perturb the orbits of Earth satellites that must be computed and accounted for many satellites.

The range of the heights of ocean tides at various shorelines can be from zero to several meters (feet). When a tidal wave piles up against certain shore-lines/ocean floor formations from certain directions, the waves can become quite high. A good example of this is the Bay of Fundy, where tides can range up to 44 feet (16 meters).

Tsunami waves behave in a similar manner. For instance, the December 26, 2004 tsunami wave piled up approximately 10-meter (30 feet) waves on many shores from Indonesia and Thailand to Sri Lanka, India and Somalia in Africa. But when traveling at jet-plane speeds over open ocean areas, the height of that tsunami wave/(waves) was/(were) only approximately one meter (3 feet) high. When that wave passed the small Diego Garcia island in the Southern Indian Ocean, the sea levels rose only by about that one meter (3 feet) because the ocean floor variations were 'too small' to prevent the free passage of the one meter (3 foot) tsunami wave without much larger wave pile-ups.

Google: < noaa's national ocean service. tides and water levels >
          < tides at bay of fundy >
          < tides of solid earth >
          < tides online >
          < ocean tides >
          < amplitude of ocean tides >
          < ocean loading tides >

< **perturbations of satellite orbits** >
< **usgs earthquake hazards program-latest earthquakes** >
< **diego garcia** >

Recall from earlier pages that **"the Moon orbits (revolves) around this Earth, and the Earth orbits (revolves) around the Sun".** **Actually, it is the barycenter** (= the common center of mass of the Earth-Moon system) **of the Earth-Moon system that orbits 'our' Sun along the pretty annual elliptical Keplerian solar orbit,** if the planetary/other perturbations are omitted. Also, the Moon's monthly orbit around the Earth is around the barycenter of the Earth-Moon system with external perturbations, which are outside the scope of this book.

Google: < **what is a barycenter?** >
             < **archive of astronomy questions** >

47.2. WHERE IS THE BARYCENTER OF THE EARTH-MOON SYSTEM LOCATED?

The barycenter of the Earth–Moon system is a point inside the Earth on the straight line connecting the centers of the Earth and Moon.

At the time of a full moon, Earth's center is inside the elliptical Keplerian solar orbit and at the time of a new moon, Earth's center is outside Earth's elliptical Keplerian orbit. The point of the barycenter 'orbits' with the Moon's monthly orbital period underground at a depth of approximately 1,700 kilometers (1,060 miles) below Earth's surface. Because Earth's radius is 6,371 kilometers (3,959 miles), the barycenter is approximately 6,371 − 1,700 = 4,670 kilometers (2,900 miles) from Earth's center, i.e. Earth's center wiggles about 4,670 kilometers (2,900 miles) in and out of the mathematical Keplerian orbit ellipse making about 12 of these wiggly corkscrew patterns every year.

The Earth and Moon are 'held together' by their mutual gravitational attractions (pulls), and they are simultaneously 'kept apart' from the barycenter by an equal but opposite centrifugal acceleration produced by their individual revolutions (orbiting) around their common center-of-mass, their barycenter. The rotational period is the same as one lunar month. One could compare this to a sledgehammer being thrown to spin along an ice rink surface, where the hammer part depicts the Earth and the tip of the handle depicts the Moon. The rotation of the sledgehammer is around the barycenter of the hammer and its handle, which is near the metal part of the hammer.

Recall during the maximum tides, the effects of the Sun and Moon on Earth's gravity are plus-minus 0.16 milligals and approximately 0.026 seconds of arc on the deflections of the plumb line, and that the tidal effects by the Moon are twice as large as those by the Sun here on Earth.

Google: < noaa national ocean service: animation of spring and neap tides >
            < the shape of the oceans >
            < ride the tide >
            < coriolis effects on tides >

Similar **differential** tide-producing forces exist also in the Earth-Sun-system with its own barycenter. It is a wobbly situation!

As the Earth turns, the small variations in tidal pulls of various areas (parts) **of Earth are in various directions** (always toward the attracting body: the Moon or Sun) during all hours of every day. These small variations are strong enough to produce the mighty daily ocean tides, Earth tides and atmospheric tides.

In ocean areas there are some so-called **nodal points**, where the tides are zero as the ocean waters slosh around this way and that way.

Atmospheric tides can be measured as small systematic changes in the atmospheric pressures. At the mid-latitudes the air tides are only around 10 microbars or 0.01 millibars. These tiny variations are almost negligible when compared to standard, average atmospheric pressure of 1013 millibars or 760 millimeters (29.92 inches) mercury.

Google: < tides and the earth's rotation >
            < tides in bay of fundy >
            < solid earth tides >
            < tides in the atmosphere >
            < atmospheric pressure >
            < nodal points of tides >

48. The Space around Us

The Milky Way, the home galaxy for this solar system, is a pinwheel shaped spiral disk galaxy. It is just one galaxy among billions of galaxies, which are all around us up to 13.7 billion light-year distances (2007) in all directions.

Reviewing our location in surrounding space, **the diameter of the disk of this (spiral pinwheel) Milky Way Galaxy with its spiral arms is approximately 100,000 light-years.** Its central bulge is approximately 6,000 light-years thick, and its thickness here near us ('our' Sun) is approximately 1,000 to 2,000 light-years. 'Our' Sun is approximately 50 light-years on the north side (North Celestial Pole and Polaris are 'that-a-way') of the central galactic plane.

'Our' Sun is at approximately **a 25,000 – 28,000 light-year distance from the galactic center, which contains** at least one **very massive Black Hole. Earth orbits 'our' Sun at a 500 light-second = at one Astronomical Unit distance.** There are 63,000 Astronomical Units in one light-year distance. **The nearest star to us next to 'our' Sun is at 4.2 light-year = 269,000 Astronomical Unit distance.**

Google: < black hole >

In space, there are no up or down directions as such. It is equally correct to say that we in America, and not the Australians, are under this Earth. **Year after year Earth orbits 'our' Sun at a speed of 20 miles (30 kilometers) per second in a very stable orbital plane = the plane of ecliptic = the Ecliptica.**

The north end of Earth's spin axis points to the Celestial North Pole of the Celestial Equator, which is nowadays near to 'our' North Star, Polaris. Generally, the two poles of a plane are 90 degrees away from their plane like the north and south poles of Earth are from the equatorial plane. The pole of the Ecliptic is at 23.5-degree angular distance from the pole of the Celestial Equator, because the **dihedral angle** between these two planes is also 23.5 degrees. The pole of the Ecliptic is 90 degrees away from the plane of the Ecliptic – like two poles of a plane are by definition. The Celestial Equator, being the extension of Earth's Equator, is the same plane as the Earth's Equator.

The covers of a hard-back book can **demonstrate dihedral angles**. When the book lies closed on a horizontal table, its front and back covers are parallel and the dihedral angle between the covers is zero. When opening the front cover to become vertical, the dihedral angle between the covers goes from zero to 90 degrees.

Google: < dihedral angle >
                < celestial poles >
                < our solar system in milky way galaxy >

Earth's orbital plane (Ecliptica) is slicing through the Milky Way Galaxy obliquely to the plane of Milky Way's galactic disk. **Our equatorial plane** (Celestial Equator) is a gyrating (remember the precession) plane slicing through the Milky Way Galaxy, gyrating by a range of 47 degrees (= plus-minus 23.5 degrees) with respect to the Eplictica, once around in 25,800 years.

The north side of the galactic plane is the side to which the north end of Earth's spin axis is pointing. For this reason the **density of the Milky Way stars with negative declinations (southern stars) is greater than the density of stars in the northern skies having positive declinations.** None of the stars with negative declinations are visible from Earth's North Pole because they are below North Pole's horizon, which is parallel to Earth's Equator. Similarly, none of the stars with positive declinations are visible from Earth's South Pole because they are also below the horizon of the South Pole.

Stars in any constellation are in one portion of the sky but they usually are at many different distances. Constellations give only a general direction to heavenly objects, such as 'Polaris is one star in the constellation of Ursa Minor'. The ancient astronomers apparently saw many animals and other shapes in the visible star groups when they connected 'nearby' stars with imaginary lines in 'dot-to-dot' fashion. Of course, they did not know the various distances to the stars, which in one naked-eye constellation can vary anywhere from 4.2 light-years to approximately 2,000 light-years.

Google: < star constellations >

Recall that most visible stars to the naked eyes are within 2,000 light-year distance. More distant stars and, of course, other galaxies can be seen only by using binoculars/telescopes. The ancient astronomers never saw any of those other Milky Way stars. They could have seen the Andromeda Galaxy at over 2,000,000 light-years distance as a hazy patch in the sky without really knowing what it was.

Stars in this Milky Way Galaxy can be up to 75,000 light-years away from us. If they are farther than 75,000 light-years, they are not in this Milky Way Galaxy disk. Beyond that 75,000 light-year distance there are galaxies of many shapes by the billions in all directions from us with billions of stars in many of them. This Milky Way Galaxy contains over 100 billion (2007) **stars. One of them is 'our' Sun.**

Constellations are used for general approximate directions in the skies. For instance, one may see a photograph of a galaxy millions of light-years away in the direction of the constellation Virgo, or Ursa Minor, etc. **The light from that far away galaxy, which produced the picture, came through the visible star field of the constellation Virgo, or Ursa Minor, etc. of the 2,000 light-year naked-eye celestial sphere in this Milky Way Galaxy.** Constellations give just the general area of the sky where the object is located. **When astronomers point their telescopes to a certain star or galaxy, they use the celestial coordinates, Declinations and Right Ascensions.** Just pointing to a large general area of some constellation would be useless.

Circumpolar stars at a location on Earth are above the horizon all the time describing their diurnal (daily) circles day after day. By definition the diurnal circle of a star is completely above the horizon if it is a circumpolar star. At Earth's Equator, there are no circumpolar stars. At Earth's North Pole, all stars with positive declinations are circumpolar stars. At Earth's South Pole, all stars with negative declinations are circumpolar stars.

Google: < circumpolar stars >

48.1. No One Knows Physical Locations of Heaven and Hell

Some people think that **Heaven is straight up in the skies.** If that is true, then it could be in any direction toward the sky. Every radius of this round Earth is straight up for some point on Earth. Individual plumb lines are pointing **in all radial directions** from the center of the round Earth.

Heaven is a spiritual realm/place/concept beyond unanimous human understanding. Heaven cannot be at some particular place somewhere on Earth including the clouds, hilltops, valleys and caves. **Its location would have been visited and mapped if it were somewhere on Earth.**

Directly above the ground/oceans is the atmosphere with only relatively few molecules above an altitude of 300 miles (500 kilometers). Above that, there is only the emptiness/vacuum of space. The next nearest heavenly objects are the Moon and planets Venus, Mars and so on. **There are just no other physical objects in our 'neighborhood' where Heaven and Hell could be located.**

All solar planets and all their moons are so well known that Heaven and Hell cannot be anywhere in this solar system. Further out, the nearest star is over four light-years away.

God is an Infinite Supreme Being. Many people believe that he is the Creator of everything in the universe. If it was not God, who, or what was it? Some believe that God is everywhere at all times. Whatever your belief is, keep it and try to keep improving your own understanding.

This book cannot provide answers to the question on Heaven's location. Just be thankful that you are here on this wonderful, unique planet Earth. This Earth is God's garden spot for you and me in this solar system.

Then where is Hell? This book does not have an answer to that question.

Heaven and Hell are spiritual quantities subject to beliefs/speculations in many religions. Many speculators think that nobody but they themselves have it figured out correctly. Many other religions disagree, and the result is that nobody knows it all. Heaven and Hell may be also in another dimension, which we humans cannot understand with our senses/reasoning. **We understand three or four dimensions pretty well, but how about ten dimensions?**

Google: < how many dimensions are there? >

There are plenty of examples of self-evident limits of all human brainpower and understanding. For example, among the many items a two-year old child cannot understand are the imaginary numbers, negative numbers or the difference in value between a one-dollar and one-hundred -dollar bill.

Experts in many fields of science often have to conclude that they do not know everything they would like know even in their own field of expertise. **The desire to learn more is a part of the reason that keeps them going!** The same is a good a goal for all of us. Fortunately, we can keep learning more about many things, but we can never learn or understand everything in the universe.

There will always be much more to learn about this wonderful planet Earth with its marvelously balanced equilibrium systems full with ingenious and complex creations, designs and natural developments. Our opportunities are endless in learning, work and in utilizing what this Earth has to offer. Humans of all religions and opinions are truly lucky to be here on this unique planet!

49. ENCOUNTERED VELOCITIES IN THE UNIVERSE

Many drive their cars 24,855 miles (40,000 kilometers) in a year (or two) **with an average speed of 50 MPH (80 km/h)** driving for a total of 500 (or 250) hours per year. Some people travel those 24,855 miles (40,000 kilometers) in airplanes in a couple of days with an average speed of 500 MPH (800 km/h). **Shuttle and International Space Station astronauts** travel around the Earth in approximately 90 minutes **at a speed of approximately 7 kilometers (4.3 miles) per second**, which speed is about Mach 21, or 21 times the speed of sound in the air at sea level.

Google: < mach number >
  < speed of sound >

http://www.grc.nasa.gov/WWW/K-12/airplane/sound.html

For comparison, the **muzzle velocity of a bullet from a M-16 military rifle is approximately 3,250 feet (one kilometer) per second**, which is approximately a velocity of Mach 3, or three times the velocity of sound in the air at sea level.

This **Earth travels around 'our' Sun at 20 miles (30 kilometers) per second**, or 30 times as fast as the rifle bullet.

Some stars near the center of 'our' Milky Way Galaxy travel at tremendous speeds of 5,000 kilometers (3,000 miles) per second, which is 'only' 1.7% of the velocity of light in a vacuum.

Together with the Sun and with some 'neighboring' stars, **Earth travels aproximately 250 kilometers (150 miles) per second around the center of the Milky Way Galaxy.**

As has been mentioned, ***the measured* distance** between the Milky Way and Andromeda Galaxies is decreasing at a rate of approximately **100 kilometers (60 miles) per second.** *However, the whole Andromeda and Milky Way Galaxies are* *decreasing their mutual distance by approximately 300 kilometers (200 miles) per second.* The explanation is as follows: In the Milky Way Galaxy our solar system is orbiting (right now) tangentially away from *Andromeda Galaxy* at approximately 200 kilometers (120 miles) per second, therefore, **the measured relative velocity** from the Earth to Andromeda Galaxy is only 100 kilometers (60 miles) per second. A completely another matter is to decide what the absolute velocities of these two galaxies are with respect to some other galaxies, or groups of galaxies. Then what about those other groups of galaxies? Where are they going? They all are moving in some direction. According to the theory of the expanding universe, most of the distances between galaxies are increasing, although there are some galaxies, which are in a process of colliding.

Everything in space is in motion. If there would be a stationary object anywhere in the space/universe, **it is still in the gravitational fields of some other 'nearby' objects/stars/galaxies. As a result, it would start 'falling' and accelerating in the direction of the resultant of all prevailing gravitational pulls (attractions) at the location of that object according to Newton's and Kepler's Laws.** The **stationary object situation does not exist after the movement starts.** So, every object in space/universe/cosmos is moving somewhere, and all objects in space are orbiting something.

For instance, the Leonid meteor shower pieces don't just sit there in Earth's 'way' every November. The Leonid debris/pieces themselves orbit 'our' Sun like everything else in this solar system, and once again next November there is another bunch of Leonid-pieces for this Earth and 'our' Moon to plow through.

An object parked on a moving object may be called stationary, such as some item on a table, a car in a garage or some instrument on the Moon. However, keep in mind that the table, the car in the garage and the Moon in the sky are taking part in Earth's motions, including our orbital speed of 20 miles (30 kilometers) per second. The Moon's orbital velocity around the Earth is approximately 1 kilometer (0.6 miles) per second.

Google: < velocity of light >
         < mach number >
         < black holes >
         < newton's three laws of motion >

< kepler's laws >
< moon fact sheet >

One gets a good idea of the order of magnitude of the great velocity of light by remembering that the distance once around Earth's equator is 40,000 kilometers (24,855 miles), and that the **velocity of light** approximately **300,000 kilometers per second equals to a distance of going 7.5 times around the Earth's Equator in one second.** This also gives a good appreciation on the accuracy of the atomic clocks, which can measure relatively easily the time it takes light to travel one millimeter distance.

50. More about Earth's Solar Orbit

Recall again that **the Earth travels** year after year in its annual orbit (one round trip around the Sun in one year) **in a very stable plane** with respect to 'our' Sun. The orbital plane is the ecliptic plane (= Ecliptica). There is no appreciable drag or friction in the vacuum of space as Earth 'zips along' with its average **velocity of 30 kilometers or 19 miles per second. At the same time, in every 24 hours of** *Universal Coordinated Time (UTC),* **the Earth spins around its spin axis** approximately **361 degrees** and not only 360 degrees, *as it does in one sidereal day.*

Earth's average distance from the Sun is **149,600,000 kilometers = 92,557,000 miles = 1 AU = one Astronomical Unit = 500 light-seconds.** Earth's orbit is slightly elliptic (it is almost a circle) with a very small eccentricity of 0.01671. Eccentricity of a perfect circle is exactly zero.

Google: < the ellipse >
            < eccentricity of an ellipse >
            < ellipse calculator >
            < earth's orbital plane >
            < the nine planets >
            < solar system overview >
            < usno seasons and the earth's orbit >

Earth's orbital plane, the plane of the ecliptic (Ecliptica), is tilted by an angle of approximately **23.5 degrees from the plane of Earth's Equator.** These two planes form a 23.5-degree dihedral angle. To visualize such an angle, think of opening a book with hard covers. The book covers form a variable dihedral angle.

Google: < dihedral angle between two planes >

The intersection of these two planes is along a straight line (back of a book) connecting two **diametrically opposite orbital points called Vernal Equinox (= first point of Aries) and Autumnal Equinox (= first point of Libra).** Along the straight line from the Earth to Sun around March 21, the Sun is in the direction of/to the first point of Aries on the celestial sphere; around September 21, the Sun is in the direction of/to the Autumnal Equinox on the celestial sphere. At these two instances, Earth is at the diametrically opposite points from the Sun, respectively.

Since Earth's orbit is almost a circle with its small eccentricity, its two foci are not very far apart (relatively speaking) from the center of the orbit ellipse. Also, consider/remember that 'our' Sun occupies one of the two foci of Earth's orbit ellipse according to Kepler's First Law. 'Our' Sun is near the halfway-point between the two equinoxial *orbital points*. For that kind of illustration, the *two orbital*, equinoxial points are separated by approximately two Astronomical Units.

Google: < the ellipse >
         < the nine planets > Scroll down to *Solar System Overview*..

Once more, at the time of Vernal Equinox around March 21 the center of the Sun, as seen from Earth, seems to cover (it eclipses it) the point of Vernal Equinox in Earth's orbit. At that time Earth itself is diametrically opposite in its orbit at the point of Autumnal Equinox.

At Autumnal Equinox around September 21 the center of the Sun (at 1 AU distance from Earth), as seen from Earth, seems to cover the point of Autumnal Equinox in Earth's orbit. At that time Earth itself is at the diametrically-opposite orbital point, the Vernal Equinox.

Approximately three months from the equinoxial points (i.e. Spring and Autumn) in Earth's orbit are the points of Solstices (i.e. Winter and Summer).

Google: < usno earth's seasons >
         < nmm equinoxes and solstices >

At Winter Solstice (Northern Hemisphere) **around December 21** the Sun's declination is approximately -23.5 degrees, and Earth is approaching its nearest orbital point to the Sun, it's Perihelion point. Then the north end of Earth's spin axis is tilted away from the Sun as far as it ever tilts.

At Summer Solstice (Northern Hemisphere) **around June 21 the** Sun's declination is approximately +23.5 degrees, and Earth is approaching its most distant point from the Sun, it's Aphelion point. The north end of Earth's spin axis is tilted toward the Sun as far as it ever tilts.

Of the Sun's nine (8?) planets Mercury is the nearest one to the Sun at 0.38 AU, and Pluto is the most distant at 39.5 AU distance from the Sun. All nine solar planets orbit in almost the same orbital plane as Earth does, which is the Ecliptica. In that respect the solar system has most of its mass in this orbital disk. Many comets also orbit nearly in this same plane, but some other comets may orbit in planes, which are almost perpendicular to the Ecliptic Plane.

Google: < **jpl solar system dynamics** >
          < **the nine planets** > JPL stands for Jet Propulsion Laboratory,
          California Institute of Technology.

If one would take an imaginary helicopter ride high above Earth's North Pole and look down on Earth as it travels in its annual orbit, one would see that the **daily spinning (rotation) of Earth is counter-clockwise, and see that the Moon orbits Earth also in counter-clockwise manner with its orbital speed of** approximately **one kilometer (0.6 miles) per second. One would also see that Earth is orbiting the Sun in a counter-clockwise manner**, as do all the other solar planets. Even the Sun itself spins in counter-clockwise direction. One may wonder if God is left-handed!

For **Earth's daily (diurnal) spinning** around its North-South spin axis, **the word used in scientific literature is rotation. Earth rotates around its North-South spin axis giving us nights and days, and Earth revolves around the Sun, giving us years.** Once more: the Earth **revolves** around the Sun and **rotates** around its spin axis.

In everyday common spoken language, the words rotation and revolution can have less precise meanings. RPM can mean either Revolutions Per Minute or

Rotations Per Minute. For instance, the RPM gauge in some cars shows rotations per minute of the engine crankshaft, although it is often called revolutions per minute.

The **unit of time = one second originates from the length of one day**: 1 day = 24 hours = 1,440 minutes = 86,400 seconds.

Other planets rotate (spin) at their own spin-rates. The duration of their days is not 24 Earth-hours, neither is the duration of their years 365.24 Earth days long.

There is no appreciable drag or friction slowing the Earth down in its annual laps around the Sun in the vacuum of space. If Earth's orbital velocity would slow down, the size of its orbit would shrink. Its velocity would thereby increase (by falling toward the Sun) and the annual orbital periods (years) would get shorter in duration. This book will not go further into that kind of speculation.

When comparing Earth to a fast-spinning, fast-pitched baseball, say at 100 MPH = 147 feet per second = 45 meters per second through the air, there are differences. The baseball will experience air drag or 'head winds' (higher air pressure on its front side than on its back side), and it will slow down. From the spinning, there are also different air pressures on the 'equatorial' sides of the ball resulting in a curveball. But if a baseball is thrown in the prevailing vacuum on the Moon's surface, there would be no air friction slowing it down because the Moon does not have an atmosphere. A baseball thrown on the Moon will maintain its 100 MPH velocity until it hits something. Curve balls thrown on the Moon would fly straight in accordance to Kepler's Laws no matter how fast they would spin. Therefore, the spinning Earth is not a curveball in its orbit.

When Earth travels in its orbit in the vacuum of space, it maintains its speed year after year according to the Law of Inertia, whose existence was first published/reported by Isaac Newton (1643-1727). Earth's atmosphere travels with Earth. **Earth's gravitation is strong enough to keep the atmosphere hugging the ground.**

Under normal atmospheric conditions of 0 C (=32 F) and 1 atm pressure, one cubic meter (35 cubic feet, 1.3 cubic yards) of air weighs approximately 1.29 kilograms = 2.8 pounds. Such air in a 12 ft x18 ft x 10 ft = 2160 cubic feet = 4 m x 6 m x 3 m = 72 cubic meter room weighs approximately 93 kg = 205 pounds.

One half of Earth's atmospheric mass is under 5.6 kilometer (3.5 miles = 18,400 ft.) altitude; the rest extends up to 300 miles (480 kilometers), and maybe on the dayside there are some air molecules even at 700 mile (1,100 kilometer) altitude. There is no exact altitude where the atmosphere ends. Of course, it is possible that some air molecules may get lost into the surrounding space. Volcanoes and some ground areas, such as some swamps, etc., put some new gas molecules in the atmosphere constantly. It is reasonable to assume that the mass of Earth's atmosphere is fairly constant.

Just like the hydrostatic pressure in water at a 5-foot (1.5 meter) depth is one half of the pressure at a 10-foot (3 meter) depth, the **atmospheric pressure at 5.6-kilometer (3.5 mile) altitude is approximately one half of the sea level air pressure.**

Google: < standard atmospheric tables >
             < Standard Atmosphere Calculator >

If there happens to be some asteroid/comet pieces 'in the way' of the orbiting Earth with its 30 km/s (19 miles/s) orbiting velocity, they can enter the atmosphere becoming meteorites. **Asteroids/comets in *solar orbits*,** which come 'in the way' of the speeding Earth here **at 1 AU distance from the Sun, can have** their own velocities **up to a maximum** of 42 kilometers **(26 miles) per second with respect to the Sun if their orbits are highly eccentric.** They may come from almost any possible direction toward the ground. Asteroids and comets in hyperbolic solar orbits have greater speeds visiting our Sun only once.

If objects in highly eccentric orbits are on a **head-on-collision** course with the Earth, they will enter Earth's atmosphere with a speed up to **72 kilometers (45 miles) per second.**

If objects in highly eccentric orbits hit Earth in the **rear-end-collision** fashion, they will enter the atmosphere with a speed of approximately **12 kilometers (7 miles) per second.**

Any collision velocity between 72 and 12 kilometers (45 and 7 miles) per second is possible for asteroids/comets in highly eccentric *solar orbits* when colliding with Earth's atmosphere.

Objects in orbits not much larger (major axis of orbit) than Earth's orbit (one AU) have orbital velocities just somewhat higher than Earth's velocity of 30 kilometers (19 miles) per second *at one AU distance* from the Sun (where we are). When they 'dip' closer to the Sun, their speeds increase further, when they go further away from us, their speeds will slow down. They also can be incoming from any direction and inclination.

On February 1, 2003, **the Columbia Shuttle Orbiter started to disintegrate** from drag/air pressure/heat in the upper atmosphere at an estimated speed of 5 kilometers (3 miles) per second, or five times faster than the proverbial 'speeding bullet'. The shuttle was orbiting at approximately 7 kilometers (4.3 miles) per second before the de-orbiting (braking, retro) rockets slowed its speed for the intended landing. Some heavy pieces from the Shuttle floated down hitting the ground at velocities up to 200 MPH, (300 kilometers per hour) = 300 feet = 90 meters per second. Items such as pieces of paper/fabric floated down at much slower speeds.

Most of the incoming space/cosmic materials are dust and powder, but there are also many gravel-size pieces and a few bowling ball size pieces every year. Because 71% of Earth's surface is ocean, many pieces fall unnoticed in the oceans; many go unnoticed even if they land on the dayside of the Earth.

Very rarely, maybe only once/few times in 100,000 years, **there may be some boulders, large objects, etc. traveling through this Solar System from other parts of this Milky Way Galaxy.** Of course, those objects may not even come close to Earth or any other solar planet, even if they travel through this solar system disk. There is plenty of room/space between the solar planets.

Some Milky Way stars do explode, and their debris can fly for billions of years in the inter-stellar space away from the explosion in many directions. Those velocities can be on the order of 300 kilometers (200 miles) per second. **Because the Sun itself, and this Earth with it, is orbiting the center of the Milky Way Galaxy at over 200 kilometers (over 120 miles) per second, the debris from star explosions can enter this Solar System with many possible velocities up to over 500 kilometers (300 miles) per second, or faster.**

There have been recent reports on measurements indicating that 'our' solar system might be entering a denser cloud of dust and debris (in 'our' pinwheel arm of the Milky Way Galaxy) than where we have been earlier. It is likely that these cloud variations from thinner and thicker volumes have existed for eons. Their velocities/direction of travel does not need to be the same as that of 'our' Sun.

Some of these forecasts call for a prolonged and increasing number of small interstellar bits entering this Solar System already at present time producing an increasing number of meteorites possibly for the next few centuries or longer. There have been some **speculations** that earlier denser dust clouds might have been related to Earth's ice ages and even to mass extinctions. However, no serious consequences are expected from the increased number of meteorites at present time. Some sample measurements have indicated that the number of incoming particles into Earth's atmosphere have tripled during the last decade.

This is also a fresh reminder that there is no such thing as a completely empty space, although the density of space dust and pieces 'out there' is very low.

The Sun (and Earth with it) is constantly plowing through some Milky Way dust/debris clouds, where almost everything including the dust/debris clouds, are orbiting the center of the Milky Way Galaxy at velocities over 200 kilometers (over 120 miles) per second, a little faster or slower than the Sun/Earth system. The relative speeds between the pieces of the dust/debris clouds and the Sun/Earth are on the order of 20-30 kilometers (6-20 miles) per second as measured from the incoming dust/debris particles.

There is nothing special to worry about these incoming tiny meteorites! They are nothing new. The situation is much the same as it has been for many centuries. **This Earth is absolutely the safest-known place for all of us in the whole known universe.**

We are truly lucky to be living here on this Earth created for us – for whom else! If not for us, then for what else might this creation/evolution be? We all have countless good reasons for trying to be happy and confident during our short life spans here on this Earth.

All the possible effects of galactic dust on Earth are unknown. Some galactic dust has been raining down on Earth for eons. An average of about 40,000 tons (there are many other estimates) of cosmic debris falls on Earth every year, or about 110 tons per day. It remains to be seen if this 40,000 tons/year has to be increased a little bit in the near future.

Google: < main asteroid belt >
             < shuttle disaster february 1, 2003 >
             < exploding star pictures by hubble >
             < interstellar galactic dust and debris measurements >
             < meteorite types >

51. TIME

51.1. WHAT TIME IS IT? WHAT IS THE ORIGIN OF OUR TIME KEEPING?

Our civil (standard) time keeping (hours, minutes, seconds, days and nights) **is determined by Earth's spin rate with respect to 'our' Sun**. As Earth's spin rate has recently been slowing down by tiny amounts, the **civil time** (the time we keep) **is adjusted by using leap seconds to conform to Earth's spin rate.** We live (keep time) by 'our' Sun and not by 'our' Moon, any planet, nor some distant 'fixed' star or galaxy. We are used to years, months, days and nights and 24-hour clocks, and we do use them.

Duration of one apparent solar day (for instance from one high noon to the next high noon, when the center of the Sun is exactly on the meridian in the south) is the time needed for this Earth to rotate around its N –S spin axis **exactly 360 degrees with respect to the Sun** or approximately 361 degrees with respect to distant stars. Because **all solar days in one year are not of equal duration**, (they vary up to plus-minus 15 minutes), the **average (mean) length of all apparent solar days in one year is used to give the mean solar time** for the days of that year. This is basically how the standard (civil) time is determined. **Then all calendar days of the year can be of the same duration. This time is uniform to** approximately **one second per one year.**

275

51.2. UNIVERSAL COORDINATED TIME (UTC) AND LEAP SECONDS

The duration of the 24-hour **coordinated mean solar day**, which our good watches/clocks are keeping, originates from averaging Earth's rotation periods over of a number of complete years. Examples of standard times are EST (Eastern Standard Time), CST = Central, PST = Pacific, and **UTC = Universal Coordinated Time, which is for the longitude of Greenwich, England for its zero (prime) meridian.**

To keep accurate uniform time over the years, Leap Seconds are occasionally added to selected days (last day of June or last day of December) to compensate for the small irregularities in Earth's rotation rate (spinning around N-S spin axis). Recall that there are more than 31.5 million seconds in one year. The spin rate (rotation rate) of Earth has recently been slowing down by less than one second per year resulting in those occasional added Leap Seconds.

Atomic clocks help make the decisions when a Leap Second should be added to a particular day. Should Earth's rotation rate ever speed up, Leap Seconds would be subtracted from the selected days of a year, but that has not happened yet. Leap Seconds are few and far in between. **All Leap Seconds have been positive since the first one was added in 1972.** This confirms that Earth's daily rotation rate has been slowing down. There have been 22 leap seconds in the 27 years from 1972 to January 1999.

The UTC is the one that is corrected for the leap seconds. Other time zones follow the UTC; the word 'coordinated' reveals it. The order of letters in UTC originates from the French language. The atomic clocks check the uniform UTC time, which is uniform except for those rare one-second steps coming from the leap seconds. Atomic clocks run in a uniform/accurate manner to the nearest second for several thousand years. For modern daily living a good time standard is a necessity in many respects.

Standard times in various time zones on the Earth differ from the UTC by a certain number of hours or by a certain number of half hours for a few countries according to the longitude zone as agreed by the legislatures of the country in question. For instance the EST (Eastern Standard Time) zone belongs to the 75 degree W longitude, which is 5 hours behind the UTC (one hour = 15 degrees).

Our good watches/clocks keep the universal coordinated time (plus-minus the time difference from Greenwich) if the occasional Leap Seconds are properly added. The UTC (Greenwich) is the basis for all standard times in all countries.

Accurate atomic clocks contribute in very important ways to our modern daily lives. Applications, which require accurate time, can be found in many areas of science, technology, communications, navigation and trade, to mention only a few areas. Atomic clocks have made modern everyday life much easier and different from the times past. For instance, among the numerous advances in science and technology, accurate atomic time has made electronic calculators, computers, cell phones, the Internet, GPS (Global Positioning System), many scientific instruments and applications possible, more accurate than ever before.

Google: < leap seconds >

< universal coordinated time >

< systems of time >

< mean and apparent solar time >

< hour angle of vernal equinox >

< autumnal equinox >

< times solstices & equinoxes >

51.3. SIDEREAL TIME = STAR TIME

The **duration of time for Earth to rotate exactly 360 degrees is called one sidereal day. It has 24 sidereal hours. One sidereal hour has 60 sidereal minutes = 3.600 sidereal seconds.** Clocks that run at sidereal rate are keeping sidereal time (star time). Sidereal clocks gain almost 4 minutes every day when compared to standard civil time. **Sidereal time is defined as the hour angle of the Vernal Equinox. Compare** this to the 361 degrees Earth rotates in one calendar day. **Solar time is defined as the hour angle of the Sun.** Compare this to the 360 degrees Earth rotates in one calendar day with respect to the Sun.

Google: < us naval observatory >

< sidereal and solar time >

< how to find sidereal time? >
< hour angle >

Sidereal years, months and days are not kept. There are no such things. There are no sidereal almanacs. For every solar day, hour, minute or/and second, there is a corresponding sidereal time for every place on Earth. When needed it must/can be found/computed for a desired instant of UTC, EST, CST and so on.

In the past, many astronomical observatories have used stars/distant galaxies for the most uniform time determinations. At the present time, observatories find the actual Earth's rotation rates, which are compared to the atomic time standard revealing whether Earth's rotation rate has changed enough for adding/subtracting another leap second. **Atomic clocks do not 'know' how fast this Earth is spinning.**

Of the many astronomical observatories, the most famous are the Greenwich Observatory in England (founded in 1675), the U.S. Naval Observatory in Washington, D.C. (established in 1830) and the Potsdam Observatory in Germany (1879). Recall that standard timekeeping is now regulated by atomic clocks that are much more accurate than Earth's rotation rates or even its annual orbital rates measured by optical means at astronomical observatories.

Google: < astronomy institutions >
            < royal greenwich observatory at herstmonceux >
            < us naval observatory >
            < potsdam astronomical observatory >
            < greenwich mean time >
            < universal time >

51.4. Leap Years by Julian and Gregorian Calendars

One complete annual 360-degree orbital travel time of Earth's round trip around 'our' Sun determines the length of one year. This **solar year has approximately 365.24219878 mean solar days** or that many rotations around the N-S spin axis **with respect to the Sun.**

In the times of Julius Caesar (100 BC – 44 BC), it was estimated that the number of days in a year (365.24219878) was 365.25, and thus every fourth year was chosen to be a leap year. That calendar is called the Julian calendar.

The difference between the Julian calendar and the present-day calendar is 365.25 minus 365.24219878 = 0.00780122 days in one year. This difference (not known this accurately by the 'Gregorians' approximately 500 years ago) is responsible for the rule by the Gregorian calendar (Pope Gregory XIII, 1502 –1585) stating that in each 400 years omit three leap years from the rule of the Julian calendar.

Let's see, how accurate the Gregorian rule is:

By the Julian calendar, there are one hundred leap years in 400 years (400 x 0.25 = 100).

By the Gregorian rule, there will be 100 – 3 = 97 leap years in 400 years.

By the present number of days in one year = 365.24219878, there should be 400 x 0.24219878 = 96.879512 leap years in 400 years.

This number 96.879512 rounds nicely to 97. The difference 97 – 96.879512 is 2 hours 53 minutes 30 seconds in 400 years.

It is amazing that the astronomers some 500 years ago, even without telescopes, were able to device such an accurate Gregorian calendar rule to have 97 leap years in 400 years by 'going wrong' only by less than three hours in 400 years.

The first telescopes were built approximately 400 years ago. Hans Lippershey (c1570-c1619) of Holland is often credited with the invention of the telescope. However, the telescope was also invented and introduced into astronomy in 1609 by Galileo Galilei (1564-1642) himself. Galileo used his wrist-pulse for timing some of his experiments getting his approximate durations of events. How he got hours and minutes of the day is unknown to the author. Well, they had hourglasses.

Google: < leap year faq >
            < leap years >
            < galileo galilei >
            < julian date converter >
            < the gregorian calendar >
            < a walk through time > by NIST
            < history of the western calendar >

51.5. SIDEREAL YEARS AND MEAN SOLAR YEARS

In one solar year or in 365.24219878 mean solar days, there is exactly one more or 366.24219878 sidereal days. In other words, **in one solar year of 365.24219878 mean solar days, Earth makes 366.24219878 complete 360-degree rotations (spins) around its N-S spin axis.**

As this Earth travels **each day** approximately one-degree arc in its annual orbit (360 degrees in 365.24219878 days, or 2.6 million kilometers = 1.6 million miles), it has to rotate approximately 361 degrees around its spin axis with respect to 'our' Sun; for instance, from one high noon to the next high noon or from one mid-night to the next mid-night.

So, to repeat, this one-degree difference (361 – 360) comes from the fact that during one day, Earth travels approximately one degree in its annual orbit around 'our' Sun. The direction from 'here' to the Sun against the star/galaxy background changes approximately by that one degree per day, or it can also be said that the radius vector from the Sun to Earth sweeps approximately one degree orbital sector every day. Therefore, Earth has to rotate about 361 degrees between two consecutive 'high noons' or between two consecutive midnights.

This explains why the 24-hour communication/weather/etc. satellites are in *sidereal* **24-hour orbits (not in 24-hour standard time orbits) for them** to stay in the same meridian plane (or nearly so) over the Earth's Equator (or nearly so) day after day.

It could also be said, that in one solar year, the Earth makes 365.24219878 RPY (Rotations per Year around its spin axis) with respect to the Sun, but 366.24219878 RPY with respect to the background stars/galaxies.

Note once more that the difference between these two numbers (number of days in a solar year and number of days in a sidereal year) is *exactly* **one.**

51:6. SUNDIALS

Each (apparent) solar day is the duration of time from one midnight to the next midnight. Midnights are used because the instant of midnight is the beginning of the next 24-hour day. High noon is the instant, when the center of the Sun crosses the meridian, or when the center of the Sun is exactly in the south having an azimuth of 180 degrees.

The **mean Sun is an imaginary Sun** seemingly 'traveling' at a uniform speed in the equatorial plane (not in the plane of Ecliptica). The actual real Sun, the so-called Apparent Sun, actually 'travels' at non-uniform speed in the plane of Earth's orbit (Ecliptica). During one year, the actual Sun may cross the meridian approximately 15 minutes before or after the imaginary mean Sun. **This is the reason time shown by the sundials can be off by plus-minus 15 minutes from the Local Mean Solar time** unless corrected for the **Equation of Time (the plus-minus 15 minutes).** By applying that correction, local mean time can be obtained from the sundial time.

In addition, to get the standard time by a sundial, the longitude difference from the standard meridian of the time zone must be taken into account. Every meridian has its own local apparent time. If one stands on the ground and faces south, the left foot has a more easterly longitude than the right foot. Time zones are 15 degrees or one hour wide in longitude. OK, it is much easier to use a wristwatch to tell the time of day; a sundial can still be a nice decoration.

During one year, the length of individual apparent solar days differ by the amount of the so called, **Equation of Time** from the duration of the Mean Solar Days. **We certainly don't want our watches and timepieces to run at variable speeds during all consecutive days of the year** varying during one year by plus-minus 15 minutes. Many inexpensive digital wristwatches run so uniformly, that the duration of two consecutive days are the same within one second in the daily 86,400 seconds.

Google: < sundials >

   < equation of time >

Sundials show the Local Apparent Solar Time, which is not uniform. Their time keeping is for the locality where they are parked. If they are positioned exactly on some standard meridian of the 15 degree time zones, such as at 75 W, 90 W, 105 W, etc., no correction for the longitude difference from the standard meridian need to be made. The Equation of Time correction must still be made to the sundial time if the standard time is desired from a sundial. Of course, sundials are not used for serious timekeeping.

51.7. Reasons for the Existence of the Equation of Time

One main reason for the existence of the Equation of Time is that the **Earth travels (according to Kepler's Laws)** *slightly faster in its orbit in the northern winters than in the northern summers*. Earth has its maximum **orbital velocity** at its Perihelion (nearest point to the Sun) in early January covering a slightly longer arc of its orbit ellipse **during one winter day than it does in early July during one summer day**, i.e., in the early July, the Earth has its minimum orbital velocity at its Aphelion (most distant point from the Sun).

The daily rotation rate around the North-South spin axis is more uniform than Earth's orbital velocity. The rotation rate is **almost** independent of annual seasons. For this reason in northern winters Earth must spin a little more from high noon to the next high noon than in northern summers. In northern summers Earth travels a slightly shorter portion of its orbit from one high noon to the next high noon than in northern winters.

Google: < kepler's laws >

The second main reason for the existence of the Equation Of Time is the fact that the Earth's **North – South spin axis** is not perpendicular to the plane of Earth's orbital plane, but it is **tilted by an angle of 66.5 degrees** to the plane of Earth's orbital plane (= plane of the Ecliptic or Ecliptica). Most Earth globes show this very clearly. Fast spinning, gyrating and wandering toy-tops on a smooth table might bring the situation in mind.

Google: < brief history of gyroscopes >

The dihedral angle between the two planes, the Equator and the Ecliptic is 23.5 degrees. At Summer Solstice in June (Northern Summer and Winter in the Southern Hemisphere), the declination of the Sun is approximately +23.5 degrees. At Winter Solstice (for Northern Hemisphere), the declination of the Sun is approximately −23.5 degrees. **Note that Earth's spin axis is perpendicular to the plane of the Equator, but the Earth orbits in the plane of the Ecliptic.** This angle of 23.5 degrees is called the **Obliquity of the Ecliptic.**

Google: < solstices >
                < declination of sun >
                < path of the sun, the ecliptic >
                < obliquity of the ecliptic >
                < dihedral angle between two planes >

52. Planet Earth, Our Sweet Home

This planet Earth is an outstanding beauty. It seems to be God's garden spot in this solar system. **There is nothing else in this solar system even coming close to it.** There may be other livable planets orbiting some stars (their Suns) many light-years away, but even there nothing similar has been found so far (2007). The search for extra solar planets continues.

This **Earth is infinitely complex in countless ways.** Its ingenious systems are in a wonderful balance in numerous ways making life possible. **God/Creator created a very good and a marvelous planet for us to live on. Feelings of awe and gratitude and the highest respect to our Creator/God should be natural for all of us.** This planet is right under our feet everywhere, whether we are in America or 'down-under' in Australia. The opportunity to live here on Earth is a great privilege, considering 'everything' here on Earth. Even the best 'experts' would have a hard or an impossible time devising anything better.

The **Moon astronauts have had a good opportunity to see and admire the beauty of this planet.** By the end of year 2003, approximately 900 passengers have orbited this Earth in space vehicles, most of them in Shuttles. Millions have seen portions of Earth from airplanes. The beauty of this Earth can be seen everywhere in its own way, even in the Badlands National Park in South Dakota.

The late astronaut Kalpana Chawla (1961-2003) describes an astronaut's view of planet Earth as seen from a Shuttle. Her description can be read at:

Google: < **UTA Alumni Astronauts – Kalpana Chawla** >
               < **photographs of earth taken by moon astronauts** >

Here on Earth, within the Earth's atmosphere, most of us have countless possibilities and opportunities to live and enjoy our lives. **There is absolutely nothing else like this Earth in this solar system.** This is the best and the only possible planet for all humans.

Other solar systems (life supporting planet/planets orbiting some suitable star) will be eternally and totally out of our reach. Space ship travel times would be several centuries long. Even a two-way radio contact (at the speed of light) with the nearest possible livable planet would take a minimum of a couple dozen years. There are no stars closer than 8 light-years from us, which could support human-type life.

Google: < nearest stars >
            < seti >

52.1. REASONS WHY THIS EARTH IS A HABITABLE PLACE FOR HUMANS

First of all, our almost constant distance to the Sun is very important. It is the Sun that maintains all life here on Earth. Without the Sun all human life here on Earth would grind to a rather sudden stop and oblivion. Earth's distance from Sun is about an ideal distance for life on Earth. There are myriads of things/ systems in this solar system, which God/Creator has made 'just about right' for humans living here on this planet Earth.

The livable zone in this solar system is 'right here' between planets Venus and Mars. Venus at 0.72 AU from the Sun is much too hot, and Mars at 1.52 AU from the Sun begins to be too cold for human life. Short duration visits to Mars are possible, but there is no water to drink, no food to eat and no air to breath.

God really created a masterpiece planet for all of us and maybe a garden spot for himself. Of course, humans cannot know about that.

There is very little that humans can do about this planet Earth, and there is absolutely nothing we can do about 'our' Sun. We cannot tweak/adjust the Sun's brightness, its eleven-year cycle, the other solar cycles, solar flares, or the fact that during its radiation storms, the Sun is spewing out billions of tons of charged gas and elementary atomic particles into its surrounding space. Some of these charged particles travel at speeds up to 1.000 – 2,200 kilometers (600 – 1,300 miles) per second reaching the Earth and the Moon (as well as all other planets) causing occasional damage to orbiting spacecraft, disrupting some power distribution grids, causing short circuits and computer reboots. Those proton storms can penetrate the skin of space suits and make the Moon astronauts sick. We can be partially prepared to deal with them when they arrive here after some 20 hours after they are seen coming. "Our' Sun operates independently and totally without any human input.

Earth's magnetic field diverts away, and Earth's atmosphere filters down dangerous parts of the Sun's radiation and reduces the intensity of the incoming solar winds composed of many elementary atomic particles. The Sun's radiation has and has had many benefits. The radiation situation for us could not be much better. Solar radiation is monitored nowadays.

Google: **< solar wind >**

               < elementary subatomic particles >

               < spaceweather.com > With daily information with archives to 2000.

The Sun's radiation has many beneficial effects on Earth's atmosphere, oceans, ground, rainfall, vegetation and human health in general. Too much sunshine can be destructive for all life on Earth. Moderation is the 'word', and the Creator made it just 'so' for us. **Please be grateful!**

The distance to 'our' Sun (one of our Creator's stars) and the intensity of Sun's radiation are just right for us. The intensity and the wavelengths of the Sun's total radiation and the distance to the Sun (one Astronomical Unit or 500 light-seconds) are about ideal in many respects. We have a Goldilocks-type climate, which considering the whole Earth is 'not too hot' and 'not too cold' for human life to prosper here on Earth. It is just about right.

The other solar planets provide good examples of these facts. It is much too hot for all life forms on Mercury and Venus, and it is mostly freezing cold on Mars, on Jupiter's moons and on the rest of the outer solar planets and moons.

Jupiter is a 'bottomless' gas planet with no visible solid surface at its non-existent 'ground level'. Saturn, Uranus, Neptune and Pluto are not worthy of mentioning in this connection.

52.2. COMPARING EARTH TO MERCURY

By the Inverse Square Law applied to planetary distances (Earth at 1 AU, Mercury at 0.387 AU from the Sun), **planet Mercury receives solar radiation, which is** approximately **7 times hotter and stronger than what Earth receives** (square the ratio 1/0.387 obtaining 6.68). **The thin atmosphere of Mercury does not filter much of the incoming Sun's radiation, nor does it moderate the temperature variations like Earth's atmosphere does for the night and day sides of this planet. The night side of Mercury is very cold.** Mercury has a huge range in temperatures. Its surface ranges in temperature from -270°F to +800°F (-168°C to +427°C)

Google: < table of planets >
            < planet mercury >
            < orbital parameters of planet mercury >
            < mercury statistics >

Among the **examples of the impossibility of any life on Mercury** are the following:

1. The **mean orbital velocity of Mercury around the Sun is** approximately **48 kilometers (30 miles) per second,** and the incoming comet/asteroid pieces are traveling in the vicinity of Mercury at speeds up to 68 kilometers (42 miles) per second. **With practically no atmosphere** (about 1% of Earth's atmospheric sea level pressure) **the comet/asteroid pieces are hitting Mercury with lethal speeds of up to 120 kilometers (75 miles) per second with kinetic energies 10,000 times greater than similar pieces would have if fired from military rifles at one kilometer (0.6 miles) per second.** Mercury's surface is full of large and small bombardment craters. The incoming objects are further speeded up a bit by Mercury's gravitational pull.

2. **The average daytime temperature of Mercury's surface is** approximately **900 F (500 C) and the average nighttime temperature is about 300 F (150 C).** Because a vacuum itself (the atmo-

spheric pressure there is pretty close to zero) does not really have a meaningful temperature, the given temperatures on Mercury are the ground temperatures. In shady areas the ground temperatures can be much colder.

52.3. Comparing Earth to Venus

1. The surface atmospheric pressure on Venus is a lethally crushing 92 bars, or approximately 90 times as large as the sea level atmospheric pressure on Earth. The same crushing pressure prevails at a depth of 900 meters (2,950 feet = 490 fathoms) in any body of water on Earth.

2. Average temperature on Venus is 850 F (460 C = 740 K). It is hot enough to melt lead.

If this Earth would be orbiting 'our' Sun at 0.723 AU distance (= distance of Venus from the Sun), or if it would receive the same intensity of solar radiation as Venus does (1.9 times stronger than Earth receives here now at 1 AU distance from the Sun), ocean waters on Earth would start boiling to steam in 'so many' years with many other associated effects and consequences.

After many millions of years, it is anticipated that 'our' Sun will enter its red-giant phase. In the process all Earth's ocean waters will boil to hot steam increasing the atmospheric pressure up to 275 times the present pressure. One can speculate that some of the upper parts of that new atmosphere will be lost in the space; but still, the new atmospheric pressure on Earth's surface will be on the order of one to two hundred times the present pressure. Compare this to the present atmospheric pressure on Venus, which is 90 times greater than that on Earth.

Google:　　　< planet venus >
　　　　　　　< orbital parameters of planet venus >
　　　　　　　< red giants >
　　　　　　　< venus statistics >

52.4. COMPARING EARTH TO MARS

1. **The average annual temperature on Mars** is approximately -100 F
   (= − 73 C = 200 K). The average daily temperatures range from −128
   F (= −89 C=184 K) to −24 F (= -31 C = 242 K). Temperatures on Mars
   rise above +32 F = 0 C only very seldom. If there is water on Mars, it
   is very cold ice.

2. **The atmospheric pressure on Mars** is from 7 to 9 millibars or approximately 0.07% to 0.09% of Earth's sea level atmospheric pressure.
   The thin atmospheric gas on Mars is not breathable, and it provides
   much less moderation/protection to the incoming cosmic rays than
   what Earth does. The gravity on Mars is too weak to keep a much
   denser atmosphere hugging/clinging to its ground.

3. The mean-orbital velocity of Mars around the Sun is approximately 24
   kilometers (15 miles) per second. The incoming solar comet/asteroid
   pieces can travel in the neighborhood of Mars at velocities up to 34
   kilometers (21 miles) per second. **Collision speeds can be up to 58
   kilometers (36 miles) per second.** The incoming pieces do not slow
   down very much in the thin atmosphere of Mars, and the Mars' gravitation tries to speed them up a bit. There is a 'sandblasting' danger for
   all equipment and visitors on Mars.

4. Similar to Earth, Mars also orbits through 'the meteor shower belts'.
   The same meteor shower belts intersect the orbit of Mars just as they
   intersect Earth's orbit – maybe even more so because Mars' orbit is
   greater than Earth's orbit.

   Google: < meteor showers > The times on the sites are for Earth,
                           not for Mars, but they exist.

5. Mars does not have an appreciable magnetic field to divert/deflect
   the incoming dangerous cosmic particles or the solar wind as Earth's
   magnetic field does. The harmful solar radiation is not filtered much
   by the thin Martian atmosphere. However, due to its distance from

the Sun, Mars receives approximately one half as much of solar radiation and solar wind as the Earth does.

6. If this Earth would be orbiting 'our' Sun at 1.52 AU distance (= distance of Mars from the Sun), or if it would receive the same reduced intensity of solar radiation as Mars receives (43% of the intensity Earth receives), all ocean waters on Earth would quite likely gradually freeze down all the way to the ocean floors in 'so many' years. Serious consequences affecting human life/survival of such an event are obvious.

Google: < planet mars >
            < how far is planet mars from the sun? >
            < mars statistics >

52.5. ATMOSPHERIC MODERATION OF TEMPERATURES ON MARS

The wide swings between day and night time (Martian days) temperatures on Mars show that the **Martian atmosphere does not do very much in reducing the day and night time temperature differences.** The daily (Martian Days) temperature swings are approximately 100 F, or 55 C degrees. Swings of that size are very rare or non-existent here on Earth on a daily basis.

52.6. JUPITER, SATURN, URANUS AND PLUTO ARE NOT LIVABLE PLACES

Google: < jupiter statistics >
            < saturn statistics >
            < uranus statistics >
            < neptune statisyics >
            < pluto statistics >
            < new horizons probe >

http://www.space.com/php/multimedia/imagegallery/igviewer.
php?imgid=4162&gid=298

Planet Earth is truly at the 'Goldilocks' distance/temperature position: Not too close to the Sun and not too far from the Sun, so that this Earth is not too hot and not too cold. It is just right!

A similar 'Goldilocks' situation exists with our whole solar system in this Milky Way Galaxy. We are not too close, or not too far from the deadly Milky Way center. The stars at the outer skirts of this Milky Way don't have enough heavier atoms for good Earth-like life. This Earth has useful atoms/material for us. There is even a special 'Earth Scientist's Periodic **Table** of the **Elements**' and their ions outlining some of the necessary ingredients for us at:

Google: < earth scientist's periodic table >
          < solar system planets >
          < The Milky Way Galaxy >
http://cassfos02.ucsd.edu/public/tutorial/MW.html

53. THE SOLAR SYSTEM IN THE BALLOON MODEL

Using the 1:5,000,000 balloon model of Earth, the Moon should be placed at a distance of (384,400,000 meters divided by 5,000,000) = 77 meters (252 feet) from the center of the balloon. The diameter of the Moon in the model would be 70 centimeters (27 inches).

Using the same balloon model to include 'our' Sun, it would be at a distance of 29.9 kilometers (18.6 miles) from the balloon. The diameter of the Sun in the scale of the model would be 278 meters (812 feet).

The following is a listing of all nine solar planets in the same scale of 1:5,000,000 with distances from the Sun and their inclinations from Earth's orbital plane.

All planets, the Moon and Sun can be considered to be in almost the same plane where this Earth orbits because the inclinations of the orbits of the other planets are only a few degrees, except for the little Pluto, whose inclination is 17 degrees. The whole solar system is essentially a disk in the ecliptic plane with lots of empty space where the nine (8?) planets with their moons orbit far apart from each other.

Using the disk shaped solar system in the scale of the balloon model, the planets would be scattered in many 'compass directions' from the Sun because the planets have their own orbital periods (years). All planets would be orbiting in their slightly inclined orbital planes referred to Earth's orbital plane, the Ecliptica. Consider this model solar disk to be the horizontal plane.

- Mercury would be at a distance of 11.6 kilometers (7.2 miles) from the Sun and never more than 1.42 kilometers (0.9 miles) above or under the Ecliptica (= Earths orbital plane).

- Venus would be at a distance of 21.6 kilometers (13.4 miles) from the Sun and never more than 1.28 kilometers (0.80 miles) above or under the Ecliptica (= Earths orbital plane)

- **Earth would be at a distance of 29.9 kilometers (18.6 miles) from the Sun and always in the Ecliptica (= Earths orbital plane = horizontal plane in the model).**

- Mars would be at a distance of 45.6 kilometers (28.3 miles) from Sun and never more than 1.47 kilometers (0.92 miles) above or under the Ecliptica (= Earths orbital plane).

- Jupiter would be at a distance of 156 kilometers (97 miles) from the Sun and never more than 3.55 kilometers (2.21 miles) above or under the Ecliptica (= Earths orbital plane).

- Saturn would be at a distance of 286 kilometers (178 miles) from the Sun and never more than 12.4 kilometers (7.71 miles) above or under the Ecliptica (= Earths orbital plane).

- Uranus would be at a distance of 574 kilometers (357 miles) from the Sun and never more than 7.71 kilometers (4.79 miles) above or under the Ecliptica (= Earths orbital plane).

- Neptune would be at a distance of 901 kilometers (560 miles) from the Sun and never more than 27.8 kilometers (17.3 miles) above or under the Ecliptica (= Earths orbital plane).

- Pluto would be at a distance of 1183 kilometers (735 miles) from the Sun and never more than 365 kilometers (227 miles) above or under the Ecliptica (= Earths orbital plane).

- **The nearest star using the balloon model** (seen through constellation Centauri) **would be at a distance of 8 million kilometers (5 million miles), or about 20 times as far as 'our' actual Moon is, but not in the plane of the solar system disk.**

54. GLOBAL TEMPERATURE MODERATIONS
BY THE ATMOSPHERE

Actual temperatures on 'our' Moon provide good comparisons/estimates of the possible temperature ranges here on Earth, **if Earth did not have an atmosphere.**

Google: **< planetary information table >**
            < surface temperatures on the moon >

'Our' **atmosphere moderates temperatures** near Earth's surface from the Equator to the North and South Poles. With the atmosphere the average global temperature is approximately 60 degrees Fahrenheit, or 15 degrees Celsius.

Earth's average distance from the Sun is 500 light-seconds, and 'our' Moon's average distance from 'our' Sun is between 499 and 501 light-seconds. This shows clearly that the intensity of the Sun's radiation is approximately the same here on Earth as it is on 'our' Moon.

If Earth did not have an atmosphere, the temperatures here on Earth would be pretty much the same as the following listing shows about the Moon's temperatures.

The Moon's average (mean) daytime surface (ground) temperature is 'cooking-hot' 216 F = 103°C. The astronauts had equipment and shoes to handle this. Temperatures in many Finnish saunas are this high, or even a little higher, where many regularly tolerate those sauna dry air temperatures for about 30 minutes.

The Moon's average (mean) nighttime surface (ground) temperature is –233 F = -153°C. The astronauts did not visit the Moon's night side. Brrrr. The days and nights on the Moon are about two weeks long.

The Moon's maximum surface (ground) temperature is approximately 250 F = 123°C. Water would evaporate very quickly at those temperatures especially in the prevailing vacuum around the Moon.

294

The minimum Moon's surface (ground) temperature is –377F = -233°C during its two-week-long nights.

These numbers speak for themselves on: what the life and temperature conditions would be here on Earth without its atmosphere. Luckily the Earth has a good atmosphere with many life-supporting features. There would be no meaningful life on Earth without its atmosphere.

Because the Moon does not have any atmosphere/clouds/gases, there is no 'air' temperature over there. Due to the Sun's radiation, or the lack thereof, the Moon's surface may be hot or cold. The Sun shines 'over there' with its full-strength heating the ground and all objects on it, which are not in a shadow of something. During the lunar nights, the Sun is under the horizon; then the ground and all objects on it cool down by radiating their heat energies into the dark vacuum of the surrounding space lowering the temperatures toward the absolute zero that is 0 degrees Kelvin = -273.15 Celsius = -459.67 Fahrenheit, reaching the mentioned minimum lunar temperature values of –377F = -233C. **The lunar days and nights are** approximately **two weeks long.** Therefore, the Moon has a two-week-long heating season followed by a two-week-long cooling season for the visiting astronauts.

Because the nights are only approximately 12 hours (plus-minus a little) long here on Earth (ignoring the six month long days and nights near the poles), those extremely low lunar nighttime temperatures would not be reached on the fast rotating **airless** Earth. But even with our 12-hour nights, the **airless Earth** would be in an unheard-of and in an intolerable deep-freeze. It is the Earth's atmosphere that keeps our temperatures tolerable day and night for existing life.

Therefore, if this Earth had no atmosphere almost everything would get cooked and dried up during the days and thoroughly deep-frozen during the nights. This all **gives a good idea how lucky we are to live within Earth's atmosphere.** Be thankful that we have air to breathe!

Carbon dioxide (dry ice) freezes at –109 degrees Fahrenheit = -78 degrees Celsius.

.Google: < temperature scales, nasa >
             < earth's atmosphere >
             < dry ice >

Without a well reflecting and insulating air-tight space suit, or an air-tight building (for all Earthlings) and almost constantly working cooling or heating, plus constantly working oxygen equipment for breathing, **it would become intolerably sizzling hot** (250 F = 123°C, or so) **on the day-side (even only for a few hours) on the** *airless* **and cloudless Earth (or on the Moon as it is)** and much too cold at nights.

In only a couple of hours on the night *side of Earth without its atmosphere,* it would be colder than being packed in dry ice (frozen carbon dioxide). The same goes for all moon astronauts. Earth's atmosphere handles that kind of situations very nicely.

54.1. GLACIER ON KILIMANJARO MOUNTAIN

Good examples of what thin air can or cannot do can be seen in the glaciers of the Andes, Mount Kilimanjaro, etc., even where the daytime sunshine is fairly hot.

The 'permanent' glacier (glaciers grow and shrink) on the Kilimanjaro mountain near Earth's Equator provides a good example on the moderating effect Earth's atmosphere has on the temperatures.

Normal atmospheric effects keep the 24-hour temperature variations moderate and more equal. The top of Mount Kilimanjaro is only 210 miles (340 kilometers) south of the Equator with a maximum elevation of 19,336 feet (5895 meters) and is topped with snow/ice. For centuries this ice mass has been increasing and decreasing. At present time the mass of the Kilimanjaro ice is decreasing due to a slow long term melting.

Mountaintops in the same geographic area as Kilimanjaro with considerably lower elevations do not form permanent ice caps because the heat capacity of the denser air is capable of melting the 'recently' formed snow/ice, if any.

The reason for the 'permanent' ice on Mount Kilimanjaro is mainly due to the **low atmospheric pressure/density/heat capacity on that mountaintop. Air pressure/density 'up there' is just under one half of the sea level values**. Therefore, there is only approximately one half as much air mass (per same volume of air) to blow by (assuming the same wind velocity) as at sea level to warm/melt the ice or let it grow. Of course, the air temperatures at that altitude are rather cold. Near the sea level elevations not far from the Equator, ice does not start to form at nights because the air keeps the surface temperatures above the freezing point of water. There may be exceptions in some desert areas.

54.2. A PARTIAL LIST OF THE BENEFITS OF OUR ATMOSPHERE

1. Atmosphere is usually taken for granted, because the pressurized air is naturally available all around us everywhere, except in vacuum chambers.

2. Atmospheric air is a multi-function thin sheet of pressurized gas, consisting mainly of nitrogen and oxygen molecules. It is a spherical blanket covering the entire Earth. Normally the exhaled human air leaves the lungs containing 14% oxygen and 4.4% carbon dioxide. Normally the inhaled air contains 21% oxygen and 78% nitrogen.

3. It is essential for all human life on Earth in many ways. Humans need oxygenated air every minute of their lives.

4. It is a fly-through medium for the birds, airplanes, helicopters and blimps. Nothing like that would be possible on 'our' Moon without rocket power.

5. It keeps us comfortable in many ways.

6. It warms and cools us.

7. Internal gasoline/diesel/compressed air/hydrogen fueled combustion engines get the necessary oxygen from the air.

8. With the Sun's help, oceans and the atmosphere are self-cleaning.

9. It is one of the wonders of the world. It is unique. It does not exist in its life-supporting form around any other object (planet or moon) in this solar system, nor anywhere else within at least a ten-light-year long distance.

10. It filters much of the dangerous incoming solar and cosmic radiation to acceptable levels. (Also, Earth's magnetic field deflects the incoming dangerous radiation away from us in beneficial ways.)

11. It is the medium for sound creation and transmissions to our ears.

12. It transports the evaporated water (rain/snow) from the oceans.

13. It is something for the weathermen and our friends to talk about.

There are many good reasons to be grateful to our God/Creator! We, together with other earthly forms of life are **the fortunate users of God's good compressed air here on Earth.** It is compressed to 1 Atm pressure at the sea level! All of us need to inhale air several times every minute of our lives.

The pull of Earth's gravity (gravitation) is our (or is it God's?) perfect air compressor. It has worked for eons on 24/7/365 basis here on Earth. It is working and compressing air continuously right now without any breakdowns of any kind. It will work for eons to come!

If one goes to the Moon or Mars, pressurized air is not available without taking the necessary breathing air/oxygen along from Earth. Oxygen tanks are heavy and they have a tendency to get empty 'too soon'.

Google: **< Atmospheric Pressure: force exerted by the weight of the air >**
              < The High Altitude Medicine Guide >
              < composition of air >
              < gas pressure > and there to:

http://www.grc.nasa.gov/WWW/K-12/airplane/temptr.html

55. Earth's Interior: Crust, Mantle and Core

The total mass of this Earth is 5.972 x 10E24 kilograms = 5.972 x 10E21 metric tons.

The three major parts of Earth's interior are the Crust, Mantle and Core. Each of them has layers.

The Crust is a relatively thin spherical shell starting from the ground surface down. Its thickness varies from 0 to 80 kilometers (0 to 50 miles). Generally the crust is thinnest under the deep oceans and thickest under high mountains. In the 8.4-foot diameter 1:5,000,000 scale model, the thickness of the Crust varies from 4 to 16 millimeters (0.16 to 0.6 inches).

At the bottom of the Crust, the value of gravity is approximately 1.002 G and the pressure (= weight of the 30-kilometer = 19-mile thick Crust above) is approximately 8,000 times greater than the sea-level atmospheric pressure.

Earth's Mantle is the next layer under the Crust and extends almost half way to the center. Its thickness is approximately 2900 kilometers (1800 miles).

At the bottom of the Mantle, the value of gravity is approximately 1.33 G, and the pressure is approximately one million times greater than the sea-level atmospheric pressure.

Earth's Core occupies the volume from the bottom of the Mantle to the center. The Outer Core is liquid and the Inner Core is solid.

The fluid viscosity of the liquid Outer Core is close to that of water providing a nice cushioning for an isostatic equilibrium for all the overlying masses.

Google :   < physical geology – interior of the earth >
                        < mass of the earth >
                        < earth's interior >
                        < the interior of the earth >

< teacher page: viscosity >
< the core >
< the nine planets > Scroll down to Earth.
< usgs inside the earth > The usgs stands for US Geological Survey. USGS-site has a good picture of Earth's cross-section.

Earth's interior is divided into several more detailed layers. Those layers have distinct chemical and seismic properties. The following numbers are depths down from the mean sea level in kilometers. The numbers differ by small amounts in various Earth-models. Average densities are in tons per cubic meter. The density of water is 1.0.

KILOMETERS

0-40	Crust	Density 2.7 to 3.3
40-400	Upper mantle	Density 3.3 to 4.1
400-650	Transition region	Density 4.1 to 4.3
650-2700	Lower mantle	Density 4.3 to 5.3
2700-2890	D" layer	Density 5.3 to 5.7
2890-5150	Outer core	Density 9.7 to 15.0. Yes, there are 'jumps'!
5150-6378	Inner core	Density 16.2 to 17.9

Most of Earth's mass is in the Mantle. The following table gives values in percents of Earth's layered masses compared to Earth's total mass.

0.000085	Atmosphere
0.02	Oceans
0.44	Crust
67.70	Mantle
30.73	Outer Core
1.62	Inner Core

55.1. EARTH'S INTERIOR IN THE 1:5,000,000 SCALE BALLOON MODEL

In the 1:5,000,000 room-size Balloon Model of Earth (described earlier), the thickness of the crust is between 4 and 16 millimeters (0.16 to 0.63 inches). For another comparison, consider an apple with a diameter of 8 centimeters (3.1 inches). Proportioning and comparing the 20 to 80 kilometer (12 to 50 mile) thick Earth's crust to the skin of this apple, the thickness of the skin of the apple would be from 0.12 to 1 millimeters (0.01 to 0.04 inches) thick. Earth's crust is a relatively thin layer.

The bottom of the Mantle is 58 centimeters (23 inches) inside the balloon model.

The bottom of the Outer Core is 104 centimeters (41 inches) inside the balloon model.

The Inner Core is a 23-centimeter (9 inch) sphere around the center of the balloon model.

Google: < earth's interior >

55.2. ARCHIMEDES' LAW OF BUOYANCY

Earth's gravity pulls all material masses in/on/around Earth toward its center. **This is also the reason why Earth is round.** Gravity pulls heavier materials harder than similar volumes of lighter materials at the same location to lower elevations. Mudslides and rolling rocks are examples of this. The final result is that **the light Crust is (floats) on top, and the heavy Inner Core around Earth's center is underneath all lighter layers**. Archimedes' Law of buoyancy applies to all layers in Earth's interior just as it applies to floating icebergs in the oceans.

Google: < archimedes principle >
              < archimedes >

One example of how heavier objects work their way to lower elevations can be seen by placing a steel ball bearing into a glass half filled with sand and then shaking the glass. The ball, which has greater density than sand, will find its way to the bottom of the glass. Lighter material has a tendency to work its way to higher elevations leaving room for the heavier material to get underneath, such as cream rising to the top of non-homogenized milk.

The floating of the lower density Crust-material upon the heavier Mantle-material is similar to icebergs floating in the oceans or ice cubes in a glass of water. This floating obeys the **Archimedes' (287 BC – 212 BC) Principle or the Law of Buoyancy for floating objects.** The Earth's thick crust under high mountains presses a depression into the underlying Mantle creating a mountain root. Earth's lighter Crust with an average density of 2.67 tons per cubic meter floats on top of the heavier Mantle with a density of approximately 3.27 tons per cubic meter. In a similar manner large floating icebergs dip deeper into the water than smaller pieces of ice. **When the Archimedes Principle is applied to Earth, it must be applied to the entire spherical Earth.**

At the bottom of the Crust and on the top of the Mantle, the density of the material changes rather abruptly. This **density discontinuity layer between the Crust and the Mantle is called the Mohorovic discontinuity layer, Moho, M-discontinuity or M-layer** in honor of Andrija Mohorovicic (1857-1936).

Google:　　< andrija mohorovicic >
　　　　　　　< crystal lattice structures >
　　　　　　　< icebergs >

The density discontinuities in the inner Earth can happen for two reasons. One possibility is that the more dense material is simply heavier material having different chemical composition of different molecules. Another possibility is that basically the same material assumes a new arrangement of its molecules into a more condensed form (another crystal structure), which is called a phase transition (change) coming from high pressures and high temperatures.

The underlying layers inside this Earth are under the pressure of all the masses (weight) of the layers above. The pressure at the Crust-Mantle standard boundary depth at 30-kilometer (20 mile) depth is approximately 8,000 times greater than the atmospheric pressure at sea level, and the temperature there is approximately 1,600 degrees Fahrenheit (870 Celsius).

Most of the heaviest earthly material is in the Inner Core at/near the center of the Earth, and the lightest material (Crust, water, air) is on top. The uppermost solid parts of the Crust are the ocean floors and the continental ground surfaces. **All layers in Earth's interior are not very far from their hydrostatic equilibrium conforming to the Archimedes' principle for floating objects.**

Earth's Mantle is not made of unyielding perfectly solid material. **When there is a consistent, strong enough force/pressure pushing for a long enough time, the Mantle material will move.**

To demonstrate, some Physics textbooks use a solid-looking piece of pitch, a steel ball bearing and a hammer. First, hit an apparently solid piece of pitch with a hammer, and it will shatter like glass. Secondly, place a steel ball bearing on top of 'solid' pitch in a bucket with the same pitch and leave it there for some time. Due to the pull of gravity, the ball bearing will sink slowly all the way through the pitch, because the density of the ball bearing is greater than the density of the pitch, and because the pitch yields. The pitch will behave like a slow-moving liquid, and the ball bearing will slowly sink into the pitch. If a light wooden sphere is used, nothing much happens, the wooden ball will just float on the pitch.

The purpose of this experiment is to show that: **"When there is a consistent, strong enough force/pressure pushing for a long enough time, the solid looking Mantle material will move."**

The whole Earth is not a perfectly rigid body. Continental wandering and other movements of Earth's Crust are commonly of the order on one millimeter to five centimeters (0.05 – 2 inches) per year in horizontal/vertical directions.

The Earth tides heave the continents and ocean floors up and down over a vertical distance up to 40 centimeters (16 inches) twice every day. Recall from earlier that even in the atmospheric high-pressure areas, the ground is pressed down by small measurable amounts.

Some parts of Earth's Mantle will typically move around a few millimeters per year if it is persistently pressured strongly enough for long periods of time. One millimeter per year for one million years is one kilometer (0.6 miles)! Where the Crust is thick, its pressure on the Mantle is greater than under the low lands; therefore, the thicker Crust sinks deeper into the Mantle under the mountains creating mountain roots.

Google: **< properties of common solid materials>**

The average density of Earth's Crust (dirt, sand, rocks, etc.) is 2670 kilograms per cubic meter or approximately 2 metric tons per cubic yard. The average density of the Mantle immediately under the Crust is approximately 3270 kilograms per cubic meter or approximately 2.5 metric tons per cubic yard.

Temperatures increase with depth in Earth's interior. The temperature near the Moho-layer is approximately 1,600 degrees Fahrenheit (870 degrees Celsius), and at the bottom of the Mantle the temperature is approximately 4,000 to 6,700 degrees Fahrenheit (2,200 to 3,700 degrees Celsius). When materials get hotter, they are approaching their melting point and can become liquid. Lava flowing down the sides of a volcano is one example.

55.3. ISOSTASY = ARCHIMEDES' PRINCIPLE FOR THE INNARDS OF THE EARTH

Isostasy is that part of Earth Science that deals with Earth's lighter top layers floating on top of the heavier underlaying layers. The floating can be in isostatic equilibrium (fulfilling Archimedes' Law for floating objects), or there can be an imbalance or other forces in play slowly working toward overall equilibrium.

Google: **< archimedes >**
                < archimedes principle >

For instance, if the Earth's Crust for some reason 'is floating too deeply' into the Mantle, the ground has a tendency to rise or to uplift from its depressed levels. That situation is occurring right now in some parts of Canada and Scandinavia where the weight of a 2-3 kilometer (1 to 2 mile) thick ice sheet (layer) pressed the Crust (ground) down during the last ice age by 200 to 800 meters (yards) into the Mantle, and the ground is now slowly rising (uplifting) toward its pre-ice age elevations. After the ice sheets melted some 8,000 to 10,000 years ago, the ground was relieved from its extra load, and the ground 'over there' has been rising ever since. The early rise (uplifting) was probably on the order of some meters, yards or feet per year. Now the ground is rising (uplifting from earlier subsided levels) mostly by less than one centimeter (half an inch) per year.

Google: < land subsidence from ground-water pumping >
             < isostasy from answers.com >
             < isostasy >
             < last ice age >

The explanation for this phenomenon is as follows: As the land-supported ice load on the ground accumulated, the increased pressure on the Crust (floating on top of the Mantle) tried to push the Mantle downward. It was 'easier' for the Mantle-material to flow sideways than to be compressed down. Some of the upper Mantle-material under the new extra loads flowed outward under the Crust in some surrounding areas, maybe under the Arctic Ocean, Atlantic Ocean, Mediterranean, Black Sea, etc. uplifting the ground (Crust) a little in those large areas.

The Canadian, Greenland and the Scandinavian ice sheets were up to 3 kilometers (2 miles) thick some 10,000 years ago. Most of their melting was finished some 8,000 years ago. As mentioned, there are estimates that the maximum ground sinking in the ice sheet-areas was on the order of 200 to 800 meters (yards).

As the ice melted, the ice age extra loads disappeared and the Mantle's movable material started to flow back' to where it came from', trying to restore the pre-ice age isostatic equilibrium between the Crust and the Mantle. The ground uplift started after the ice had melted enough. The uplifting is still continuing today at a rate of a few millimeters or centimeters per year. The maximum rate in Scandinavia is now 9 millimeters per year, some areas in the Western Canada are uplifting faster, which may be due to some other tectonic causes.

It should be noted that the **ice must be land-supported** (floating ice does not count) for it to have the described the effects on land subsidences and uplifts. For instance, if the **floating ice** in the Arctic Ocean increases or decreases in mass, there will be no appreciable effects on the sea levels.

The areas of rising ground after the ice age are said to be **over-compensated**, or the root of the Crust is sticking 'too deep' into the mantle.

If an area of ground is sinking (subsiding), one reason can be that the root of the Crust into the Mantle is not 'deep enough' for isostatic equilibrium. If so, the area is **under-compensated**.

Generally, it is a rather normal situation in Physics that 'things' tend to move toward their equilibrium positions.

The normal thickness of Earth's Crust under the shorelines is approximately 30 kilometers (20 miles) in some isostatic models. The maximum thickness of the Crust is approximately 80 kilometers (50 miles). The minimum thickness of the Crust is under the deepest ocean waters where the crust can be missing completely.

Google: < isostasy >

< **major tectonic plates of the world** >

One numerical Example of Isostatic Floating of Earth's Crust on top of the Mantle follows:

Consider a 35-kilometer (22 mile) thick Earth's Crust with a density of 2.670 tons per cubic meter at some ocean coast. Then consider going inland some distance to the top of 2-kilometer (1.2 mile) high mountain ridge, which is in isostatic equilibrium floating on top of the underlaying Mantle with a density of 3.270 tons per cubic meter.

Consider drilling into the ground from the mountaintop. After drilling down 2 kilometers (1.2 miles), the tip of the drill is at sea level. Continue drilling down another 35 kilometers (22 miles) through a part of the Crust. The density of the material around the drill hole is basically the same 2.670 tons per cubic meter.

From this level (35 kilometers = 22 miles) below sea level (standard thickness of crust), consider drilling further through the mountain root for 8.9 kilometers (5.5 miles) down until starting to enter the Mantle. The density of the material around the drill hole thus far has been basically the same 2.670 tons per cubic meter all the way from the mountaintop down to the tip of the mountain root dipping 8.9 kilometers (5.5 miles) into the Mantle.

The crustal material with the mountain, 35-kilometer crust and the mountain root in this example is 2 + 35 + 8.9 = 45.9 kilometers = 28.5 miles = 43,900 meters thick under the mountain down to the Mantle.

Let's check the pressures at the depth of the tip of the mountain root and at the same depth under the shoreline assuming standard gravity of 1 G.

The pressure (metric tons per square meter) inside Earth at the tip of the mountain root is 45,900 x 2.67 = 122,600 tons per square meter = the weight of the one square meter column of 2.67 tons per every meter above. This is 12,260 times greater than the standard atmospheric pressure on the ground at sea level.

The pressure (tons per square meter) inside the Earth under the coastline *at the depth of the mountain root* is as follows: First, there is 35 kilometers of Crust to the Mantle, and then there is 8.9 kilometers of Mantle material, or 35,000 x 2.67 + 8,900 x 3.27 = 122,600 tons per square meter.

This example shows that the pressures inside this Earth at these two locations at 43.9 kilometers (27.3 miles) below sea level are equal. This was just one numerical example of isostatic equilibrium.

Similar examples could also be given for the anti-roots of the oceans by using the same Archimedes' Principle.

For comparison again, the standard atmospheric pressure (weight of the air above) at sea level is 29.92 inHg = 760 mmHg, or 10 metric tons per square meter.

The oceans have anti-roots where the crust is thinner than normal, and the Mantle thus comes up higher than normal. Over large surface areas, the isostatic floating of the Crust on the Mantle is not very far from isostatic or hydrostatic equilibrium according to Archimedes' Principle. If the root is just right ('Goldilocks-type situation'), the area is said to be in isostatic equilibrium.

Earth's surface is made of 'broken up' tectonic plates (a dozen areas of the Crust) that move around slowly in many ways. The ground is uplifting in some parts of Canada, Scandinavia, Himalayas, etc. The ground is sinking around Denmark, Holland and at some places in California, Houston, Texas, and the Atlantic Ocean is getting wider, etc.

Google: < major tectonic plates of the world >

The oceans waters on top of the ocean floor are also close to their isostatic equilibrium. **If not, there would be a tendency of the ocean floors to move up or down toward their isostatic equilibrium.** Under deep oceans the heavier Mantle starts closer to sea level than under the continents. Oceans have their taller anti–roots there. The roots (down) and the anti-roots (up) start from the average depth of the interface of the crust lying on top of the mantle.

The average water depth of Earth's oceans is 3,900 meters = 12,800 feet = 2,130 fathoms. **The water pressure at 3,900-meter water depth is 3,900 metric tons per square meter, or 390 times the atmospheric pressure at the sea level.** At 10,000 meter-depths, this pressure is 10,000 tons per square meter.

There is a depth in Earth's interior where the pressure (the weight per a square meter/foot of overlaying material) is approximately the same all around this Earth. That level is almost a spherical shell around Earth's center. Pressures at deeper, constant pressure levels are greater.

The pressures in Earth's interior are caused by Earth's gravity (= gravitational pull + centrifugal acceleration coming from Earth's daily rotation). All upper layers are pressing down on all the layers below. The pressure at Earth's center is on the order of four million times the atmospheric pressure at sea level.

The value of gravity increases by 0.004 % when going down into Earth for the first 80 kilometers (50 miles). When going deeper, the value of gravity increases further to approximately **1.089 G near the boundary of the Mantle and Outer Core. Going deeper, the gravity decreases reaching zero G at Earth's center.**

After Archimedes, Isostasy and subject matters relating to gravity have been studied/considered/published by many scientists. Among them are:

Leonardo da Vinci (1452-1519),

Pierre Bouguer (1698 – 1758),

J.H. Pratt (1800-1867),

F.R. Helmert (1843-1917),

J.B. Airy (1855),

W.A. Heiskanen (1895 – 1971),

F.A. Vening-Meinesz (1887-1966).

Type the names in Google's search engine if information about the work of these scientists is desired.

Google: **< important geoscientists >**

< isostasy >

< usgs gravity and depth to basement method >

< isostatic models-examples of the eastern alps >

< Crustal loading, isostatic models and lithospheric flexure >

Scroll down on the sites to see illustrations/maps/photographs.

56. Ground Movements Up, Down and Sideways

Recall that the relatively thin Earth's Crust is a spherical shell all around the globe. It is not very far from being in its hydrostatic/isostatic equilibrium floating on top of the Mantle. At many places the ground is slowly and continuously moving toward its isostatic and other equilibrium positions usually by a few millimeters per year for thousands and millions of years. The **ground may be sinking** (subsiding), **rising** (uplifting) or **moving horizontally**.

There are many reasons for the more or less continuous vertical and horizontal ground movements. Earth's surface is made of broken-up pieces of its Crust, so-called **tectonic plates**, which move in many ways relative to the other pieces of the tectonic plates.

Google: < plate tectonics >
           < 1964 good friday earthquake in alaska >
           < usgs earthquake hazards program-latest earthquakes >

Local earthquakes can move the ground by several yards (meters) rather quickly. For instance, the 1964 Good Friday Alaskan **earthquake changed some topographical elevations on Montague Island and some surrounding areas up to 15 meters (50 feet) in a short period of time.**

In some situations, **it does not take very large pressure changes to move the ground level up and down.** Examples of this are **high and low ocean tides on some shorelines, called ocean loading.** Even the **atmospheric pressure changes** of highs and lows cause measurable ground movements up and down.

Even on a smaller scale there are ground movements produced by **new large dams built across some rivers.** The accumulating mass of water behind the dam will press the affected ground areas down. Many water reservoirs and dams are regularly monitored for their slow sinking/movement.

Google: < ocean loading >
          < atmospheric pressure loading of ground surface >
          < dam monitoring >
          < land subsidence in the united states fact sheet >
          < Ground motion measurement in the Lake Mead area, Ne-
          vada, by...>

Also, where **large amounts of ground water are extracted, the ground may sink**. It is unlikely that such sunken areas will ever rise back to their previous elevations.

The given Internet site of US Geological Survey **"land subsidence in the united states fact sheet"** has a photograph of a telephone pole showing how much the ground level has subsided from 1925 to 1955 to 1977. The average subsidence over those 52 years has been on the order of one foot per year. The site of sinking is in the San Joaquin Valley southwest of Mendota, California. The same Internet site lists many reasons for land subsidence in many areas of the United States.

Google: < this dynamic earth >
          < historical perspective [this dynamic earth, USGS] >
          < alfred wegener, continental drift, and plate techtonics >
          < usgs >

These three US Geological Survey Internet sites deal with present and historical worldwide tectonic plate movements and also with continental wanderings.

Google: < **alfred wegener, continental drift, and plate techtonics** >, (yes, it is spelled techtonics in the Internet address) has seven **informative links about plate tectonics.**

When some of *the land-supported ice sheets* partially melt, the water from the melted ice finally runs into the oceans, making all oceans around this Earth 'so much deeper'. The world sea levels will not necessarily rise by equal amounts because the added water to the oceans will press all ocean floors down, increasing the volume of ocean basins. By the way, even the 2-3 kilometer (1.2 to 1.9 mile) thick Antarctic ice is called an ice sheet.

Approximately **99% of Earth's land-supported ice masses are in Antarctica and Greenland.** The ice-covered mountaintops here and there around the globe have only miniscule amounts of ice when compared to Antarctica and Greenland.

There are large amounts of floating ice in the Antarctic and Arctic oceans and in the floating ice-shelves around Antarctica. **Recall from earlier that when the** *floating ice* **in the oceans melts,** *there is no appreciable non-thermal change in the sea level as the result of that melting.*

If all land-supported ice **on Earth would melt, the** *ocean depths* **would increase on all oceans by** approximately **80 meters (250 feet).**

It would be wrong **to state that the** *global sea levels* **would rise by the same amount, as the media often reports.** *Repeat: The sea level would not climb up the 80-meter topographic contour line.* The Statue of Liberty near New York City would not get submerged!

The weight of the additional water would put 80 metric tons of extra pressure on every square meter of all ocean floors covering 71% of Earth's surface. Recalling that even the small atmospheric pressure changes move the ground up/down, and that the ocean floors sink and rise twice every day from the weight variations of the small daily high and low tides, it is obvious that an extra 80-meter water load would move the ocean floors even much 'easier'. Note that the total air pressure exerted upon Earth's surface by the weight of its atmosphere is approximately the same, as a 10-meter (30-foot) deep-water layer would exert on the ground In the absence of our atmosphere.

As a result of such imaginary melting of all land-supported ice, the ocean floors would sink, and if Earth's volume is to be preserved, the continents must rise. The result is that at the oceanic shorelines **the sea level would not climb up to the 80-meter elevation contour line,** as some media news stories repeatedly tend to report.

There are more than a dozen other scientific reasons for possible sea level changes in addition to the land-supported, ice-sheet melting or growth, i.e. the budgets of land-supported, ice-sheets must be known/measured/considered **before scientific and meaningful conclusions about sea level changes can be drawn.**

Google: < mean sea level variations >

The hydrostatic/isostatic equilibrium between the oceans and the continents is another beautiful example on how the Creator has balanced even the world's sea level.

57. EARTH'S OCEANS

Without water in 'our' oceans, there could be no human life on Earth. This Earth is the only planet/object in this solar system on which water exists in liquid form.

The **mass of all ocean waters is** approximately **1.4 x 10E18 metric tons.**

Oceans cover approximately **71 % of Earth's surface.**

The **mean depth of all ocean waters is 3900 meters = 12,800 feet = 2100 fathoms.**

The volume of ocean waters is 1.4 x 10E18 cubic meters = 1.4 x 10E9 cubic kilometers = 3.4 x 10E8 cubic miles, or **a volume of a cube with a side of 1,120 kilometers = a volume of a cube with a side of 695 miles. One cubic meter (1.3 cubic yards) of distilled water has a mass of** approximately **one metric ton. One cubic meter of ocean water has a mass of** approximately **1.027 metric tons.**

Google: **< geodetic and geophysical data of earth >**
             < noaa ocean explorer >

The Sun's radiation evaporates more than one-meter (three to four feet) thick layer of water off the oceans every year. The evaporated water is essentially distilled water vapor forming clouds that carry rain and snow around the globe. The evaporated water from the oceans feeds the lakes, rivers and swamps and replenishes the ground water. **Without rain/snow, most if not all lakes, rivers and ground water would finally dry out, and all land areas would become dry deserts.** Without water there would be no plant/animal life or growth.

Oceans are an integral component of Earth's weather patterns. **The mass of annual evaporated water off the oceans is** approximately **3.6 x 10E14 metric tons** = 360,000 cubic kilometers of water = 87,000 cubic miles of water, or a volume of a cube with a side of 70 kilometers = side of the cube with a side of 44 miles, or = 60,000 tons of water per person assuming the world's population to be 6 billion.

Clouds carry this tremendous annual amount of water. For example, a **four-inch** (10 centimeter) **rain over** an area of one square kilometer (0.39 square miles = **2,040 acres**) dumps **100,000 metric tons of water** down on that area. Many rainstorms cover much larger areas than 2,040 acres, and the amount of rainfall can also be much more than the mentioned four inches.

Less than one half of the incoming sunlight entering the ocean surface penetrates deeper than approximately one meter (one yard) into the water. It gets increasingly dark in the oceans as the depth increases. **Most food production by photosynthetic marine plants can take place only in the relatively thin sunlit layer of the near-surface water.**

The Sun's radiation keeps the production of photosynthetic marine plants going. A **large part of oxygen produced and released into the atmosphere is a by-product of the photosynthesis by plants and algae living in the upper layers of the oceans.**

The oceans are a part of the natural **exchange of carbon dioxide with Earth's atmosphere.**

Google: < photosynthesis in the oceans >
              < photosynthesis in green plants >
              < el nino ocean current >

Oceans, together with the green plants on land areas, deliver the re-circulated oxygen molecules into the atmosphere by photosynthesis. To survive, we all need to inhale some oxygen every minute of our lives.

Atmospheric air is an ideal gas mixture for us to breathe. The inhaled air we breathe has a little greater concentration of oxygen molecules in it than our exhaled breath. We share the same atmospheric gases with other breathing creatures. A part of the air that just went into our lungs may have been in the lungs of a dog, horse, rat, opossum, etc. just a few seconds or few minutes earlier.

Oceans are a good **source of food** for most of us. Oceans also provide an important and economical mechanism for **travel and transportation/shipping.**

Google: < physical parameters of world oceans >

               < photosynthesis >

               < el nino ocean current >

               < oxygen production in nature >

               < carbon dioxide in oceans >

               < geodetic and geophysical data of earth >

57.1. OCEAN CURRENTS

Ocean waters move (swirl) around constantly in a somewhat similar manner as the regular/irregular winds move around visible and invisible smoke, gas, flags waving, dust, clouds, rain, small and large objects. In addition to some prevalent directions of movement, the **ocean waters are also swirling and agitated by** continental-sized gyres as well as centimeter-sized turbulent eddies **where colder and warmer waters are mixed and stirred.** The movements of ocean waters influence the climate and living conditions for plants and animals all over the globe.

Ocean waters never come to rest as long as 'our' Sun shines/heats the surface waters on the dayside of the spinning Earth as it does 24/7/365.24. The Sun does it, and it has been doing it every minute of every day for millions of years.

Ocean waters are also heated to **a lesser extent by volcanic ocean floor heating at over 5,000 locations making the warmed-up waters move upwards.** The daily tidal waves mix together with the solar heated water movement causing a part of ocean water circulations. Somewhat similar water movement can be observed in a heated water container on a kitchen stove, when warmer water rises and causes other parts of the water to move around in some way. Of course, ocean waters move around in more complicated ways than water/soup in a kettle on a stovetop.

Google: < underwater-submarine volcanoes >

There are both surface (upper 400 meters/1300 feet) and deep-water currents in the oceans. The primary forces moving the ocean waters around are:

Solar and volcanic heating
Winds
Gravity, gravitations of Moon, Sun and Earth
Coriolis forces

Google: **< ocean gyres >**
             < ocean currents and climate >
             < coriolis effect >

There are dozens of large ocean currents. Of the many **large consistent ocean currents**, maybe the best known is **the Gulf Stream** flowing from the Caribbean by the American East Coast north-easterly across the Atlantic Ocean to and by the coasts of the British Isles and Scandinavia to the Arctic Ocean and then out of there. Other well-known ocean currents are the **El Niño and El Niña** in the Pacific Ocean flowing toward the coast of South America and then forced north/south/under to deeper waters by the South American coast.

Ocean waters cannot pile up very high anywhere because Earth's **gravity makes the waters always flow toward lower elevations (down hill) in oceans as well as in rivers and elsewhere.** For this reason, in the oceans there must be some return of water and constant circulation as the water has to flow somewhere from a higher elevation where it was piled up. El Niño is an example of a current piling up against the South American shorelines.

Surface ocean currents, such as the Golf Stream and the El-currents, move their primary water in the upper 400 meters (1,300 feet) of the ocean moving large amounts of water. For instance, the **Gulf Stream near Nova Scotia transports over 150 million metric tons or 150 million cubic meters or 0.15 cubic kilometer of warmer water per second through the stream's variable cross-section** from the Caribbean all the way across the Atlantic Ocean.

The Gulf Stream causes **a much warmer climate for the British Isles, Scandinavia and Northern Europe than what would exist without the Gulf Stream.** Greenland, which occupies approximately the same northern latitudes as Scandinavia, provides a good example of what the warmth of the Gulf Stream can do. Greenland has approximately the same amount of daily sunshine as Scandinavia does, but it does not have a warm ocean current warming up its climate. The differences with the Northern European climate are great. Greenland is covered up to 3,000 meters (9,800 feet) thick, land-supported ice sheet. Agriculture flourishes in many areas of Scandinavia at Greenland's latitudes.

Google: **< current velocities of the gulf stream >**
                  < el niño >
                  < ocean currents >

58. EARTH'S ATMOSPHERE

The atmosphere surrounds the oceans and the 'solid' Earth completely as a relatively thin spherical shell. Approximately **one half of the atmospheric mass is under the altitude of** approximately **5.7 kilometers (3.5 miles, or 19,000 feet).** The rest of the thinning atmosphere extends to a few hundred kilometers (miles) higher. There are still a few air molecules at the International Space Station altitude of some 400 kilometers (250 miles) and even higher. The air molecules 'in the way' slow down the Space Station (orbiting at approximately 7 kilometers = 4.3 miles per second) and other satellites ever so slightly. This small deceleration causes the so-called micro-gravity, where the on-board objects on the Shuttle have a tendency to move forward similar to objects in a car that is being braked down to a slower speed.

There are many other types of orbital perturbations.

Google: **< orbital perturbations of earth satellites >**

To see how thin Earth's atmosphere is, go to:
http://antwrp.gsfc.nasa.gov/apod/ap051207.html

Click on the image to enlarge it. Scroll the image. Recall that Earth's radius is 6371 kilometers = 3959 miles, and that there are not very many air molecules above 300 kilometers = 200 miles altitude.

The **mass of Earth's atmosphere is 5.1 x 10E15 metric tons.**

The standard atmospheric pressure at sea level (1 atm) is defined using many units. The most common ones are 29.92 inches, or 760 millimeters of mercury, 1,013.25 millibars, 14.7 pounds (6.8 kilograms) of force on every square inch, or 1.013 kilogram force on one square centimeter, or 101,325 newtons per every square meter, **or 10 metric tons per square meter**. The standard pressure is an average representative for the sea level value. The actual atmospheric pressures vary with elevation from place to place and moment to moment everywhere on Earth. Air is surrounding and pressuring all surfaces on Earth including our skin from head to toe.

The weight of the atmosphere produces a barometric gas pressure on all exposed surfaces in the atmosphere. The **atmospheric pressure is produced by Earth's gravitation,** pulling all air molecules toward lower elevations. In addition, the molecules at lower elevations are pressed further down by the weight of the molecules at higher elevations because they all 'want to' crowd down as far as possible.

Google: < earth's atmosphere >
           < atmospheric pressure >
           < air pressure on earth >
           < ams glossary >
           < atmospheric density >
           < atmospheric density dynamics and the motion of satellites >
           < encyclopedia:pascal (unit) >
           < recorded weather extremes >

The **density (mass divided by volume) of dry air** at 0 degrees Celsius = 32 Fahrenheit at a pressure of 760 millimeters mercury = 29.92 inches mercury is = 1.3 grams per liter **or 1.3 kilograms per cubic meter or** approximately **2.2 pounds per cubic yard.** Atmospheric air pressures and densities decrease with increasing elevation. Cold air is denser than warmer air.

On TV and other weather reports the atmospheric pressures are reduced to sea level instead of reporting actual air pressure. For instance, if the weather conditions happen to be about the same at a place at sea level in New York City, NY and at a spot at an altitude of 6,000 feet (1829 meters) near Denver, Colorado, the pressure at sea level in New York could be 29.92 inches (760 millimeters) mercury, and it would be so reported on the New York Weather Report. **The normal actual air pressure at that 6,000 foot altitude spot near Denver would be only 23.98 inches (609 millimeters) mercury, but the TV Weather Report in Denver would report the air pressure at that 6,000 elevation to be 29.92 inches (760 millimeters) mercury because the actual air pressure 23.98 in. (609 mm) is reduced to sea level from the 6,000 foot elevation.**

Google: < altitude sickness and atmospheric pressure >
              < understanding air pressure >
              < definition of atmospheric pressure reduced to sea level >

Actual air pressures in deep mine shafts may be 1.5 times the normal sea level pressure or 45 inches (1,150 millimeters) mercury.

Jet planes provide a good example of reduced air pressure at high altitudes. **Airplanes flying above 10,000 to 12,000 feet** (3 to 3.6 kilometers) **need air pressurization or oxygen masks** for the crew/passengers, because the air at those altitudes does not have enough oxygen for good breathing and blood oxygenation. **If an airplane loses cabin air pressure at higher altitudes, those aboard not getting their oxygen masks on quickly, could lose consciousness or even die.** Airplanes have systems and capable pilots to handle such emergencies. Rare unexpected accidents can still happen. One such lethal event happened in a Learjet flying between altitudes 22,000 and 51,000 feet (6.7 to 14 kilometers) on October 26, 1999, when all on board lost their ability to function, and either died on the plane or in the crash that followed.

Google: < october 26, 1999 learjet crash >

The air we breathe contains a large number of molecules, mostly nitrogen and oxygen molecules. In one minute or two we may inhale/exhale one mole of air in and out. The number of air molecules in a mole of air (approximately 29 liters or 7.7 gallons) under normal conditions (pressure, temperature, etc.) is equal to the Avogadro's number 6.02 x 10E23.

Google: < density of air >
              < mole concept and mole conversions >
              < nist, si system >
              < avogadro's number >
              < composition of earth's atmosphere >
              < The atmosphere – origin and structure >

In this solar system there is no other planet or any moon of any planet (Jupiter has at least 16 major moons, Saturn has at least 18 major moons) **with**

a breathable atmosphere. **On any solar system planet/moon, we could not breathe more than one or two fatal final gasps.**

In 'our' Milky Way Galaxy there may be many planets with Earth-like atmospheres orbiting their unknown and unreachable stars, which are somewhere between 10 and 70,000 light-years away (2007) from us. **Humans will never know very much about them.** It is also a total impossibility for any Aliens from those far-away worlds to ever visit this planet Earth.

Be grateful to your Creator/God for this wonderful livable and unique planet Earth! This is the place to live our lives! Here, and nowhere else, we have air to breathe. There just are no other possibilities. 'Our' Moon and planet Mars may be visited by astronauts only for short durations of time, if they bring along their own breathing-air, etc. Even Jupiter's moons may already be permanently out of human's reach for several reasons. Robots will visit those places long before humans will even try it.

The constant availability of oxygen gas for all human life is absolutely necessary. Here on Earth one can go anytime, almost anywhere and there is always breathing air for all of us. (Omit some rare local poisonous conditions and stay mostly at lower elevations than 10,000 to 15,000 feet (3,000 to 4,600 meters, or take along your own breathing-air). **We need oxygen in/out our lungs every minute of our lives.** Many of us cannot hold our breaths for 30 seconds.

58.1. Atmosphere has Many Functions

The following are among the many functions the atmosphere has:

1. The atmosphere **filters the dangerous solar and cosmic radiation** bombarding the Earth day and night. Unfiltered cosmic radiation would be lethal for humans in a few years, if not in a few months.

2. The atmosphere **provides life-supporting oxygen**, which is available for everyone at all times around the Earth.

3. The atmosphere **is the medium for sound/speech/music** to transport sounds into the ears of listeners. The air pressure changes we

hear vibrate up to twenty thousand (or more) separate sound receptors in our inner ears so that we can hear music, the spoken words and other sounds. There are no sounds around the Moon without radios. The ground might be vibrating there, but lunar astronauts must use radio communications, and their ears must be in the atmosphere of their space suits/ships. The air traveling in and out of our lungs vibrates our vocal cords to make words, music and other sounds.

4. The **atmosphere slows down small meteorites or burns them to dust** floating harmlessly down to the ground. Without air braking, a large number of meteorites (pieces of rock and ice, etc.) would pepper the Earth's surface in lethal fashion with velocities between 10 and 70 kilometers (6 and 45 miles) per second. If there was no atmosphere, the incoming meteorite pieces would be further accelerated to even greater speeds by Earth's gravitation.

Those pieces in the vacuum of space are moving at velocities from 10 to 70 times the muzzle velocity of a bullet fired from a military rifle. Without the atmospheric braking, a small meteorite pebble weighing only one percent of a rifle bullet would be more lethal than the bullet from a military rifle. It is estimated that an average amount of meteoritic mass of 1,000 tons to more than 300,000 metric tons lands on Earth every year.

Google: < 10-Accretion of Mass >
              < meteorites falling on earth >
              < meteors and meteorite showers >

5. The Earth satellites and the Moon astronauts in the vacuum of space are in more danger if there happens to be incoming comet/asteroid pieces where they happen to be. Holes in some satellite solar collector panels have been observed. Due to the high velocities, the pebbles making those holes are more dangerous than rifle bullets. **The Mars astronauts will be in similar danger due to the low density Martian atmosphere.**

6. However, the vacuum of space is very empty, and the cross-section area of a satellite is very small compared to the whole Earth, which receives over 1,000 tons of meteorite material per year. Of the hundreds of satellites between 1963 and 1998, there have been only two or three satellite failures for which meteoroid impact was identified as the most probable cause.

Google: < human inner ear >
< cosmic radiation hitting the earth >
< clouds around the earth >

58.2. FURTHER BENEFITS OF EARTH'S ATMOSPHERE

Our atmosphere is usually taken for granted, because it is naturally available wherever we are.

1. It is a multi-function thin sheet in the form of a spherical shell of oxygenated gaseous medium covering the whole Earth.

2. It is essential for all human life on Earth.

3. It is something to fly through for the birds, airplanes and helicopters. Nothing like that would be possible on the Moon without rocket power.

4. It is our breathing air within reasonable temperature ranges.

5. It warms and cools us.

6. Internal gasoline/diesel/compressed air/hydrogen fueled combustion engines require oxygen from the air to function.

7. With the help of the Sun and the oceans, the atmosphere is self-cleaning.

8. It is truly one of the wonders of the world.

9. It does not exist in its life-supporting form around any other object (planet or moon) in this solar system.

10. It is something for the weathermen and our friends to talk about.

11. **It is another good reason to be grateful to our God/Creator!** We are just the lucky users of God's good air.

Earth's atmosphere filters much of the incoming dangerous **solar and cosmic radiation. Without atmospheric filtering and without deflection by Earth's magnetic field,** this Earth would be very dangerous place over extended periods. The radiation on the Moon is as dangerous as it would be on an airless Earth. The future Moon astronauts must be coping with that fact.

59. Global Warming and Cooling

When comparing Earth's water masses and their heat capacities to air masses and their heat capacities, one sees at once that **the mass alone of the ocean waters is** approximately **275 times greater than the mass of the total atmospheric mass. The heat capacity of water is much greater than the heat capacity of atmospheric air.**

When the Global Warming/Cooling is discussed in the news media, usually only the average air temperatures are considered, and the average ocean water temperatures are ignored.

Google: http://www.clearlight.com/~mhieb/WVFossils/ice_ages.html

               http://www.physicalgeography.net/fundamentals/7x.html

               < the nine planets > Scroll down to Earth

               < greenhouse gases >

               < heat capacity >

               < heat capacity of water >

               < heat capacity of air >

               < ams glossary >

               < global oceanic heat budget >

Earth's ocean waters are much greater heat sinks than the atmospheric air masses are. Oceans have a great effect in controlling local atmospheric temperatures and global warming/cooling. The Sun itself, of course, is the original source of controlling all global temperatures. Water vapor and other greenhouse gases in the atmosphere together with the cloud cover affect/modify Earth's local and mean temperatures. Volcanic heating at some locations is also a factor, but 'our' Sun with its cycles is the major driving force for global heating and cooling.

Some water evaporated from the oceans stays as snow on Antarctica, Greenland, and in some glaciers here and there around the globe. Some glaciers are melting more than growing, and some icebergs are calving from Greenland and Antarctica. The total ice budgets of Antarctic and Greenland ice constituting approximately 99% of all ice masses on this Earth are not known very well. Hopefully satellite measurements/observations may be able to determine the world's ice budgets in the near future.

It is meaningless to draw conclusions about the global warming based on the melting of some continental icecaps, which have only 1% of Earth's ice masses, while ignoring what happens to the remaining 90% of Earth's ice in Antarctica and 9% in Greenland. Nobody knows whether the total Antarctic and Greenland ice masses are growing or shrinking or by how much.

It is also rather meaningless to talk about global warming/cooling based solely on air temperatures, when the known heat capacity of the oceans is much greater than that of the total atmosphere. When the average ocean temperatures (from top to bottom) are properly combined with the average air temperatures, better scientific information becomes available on global warming/cooling.

Google: **< heat capacity of oceans >**

There are some signs that the next ice age has already started. It remains to be 'seen' if its maximum will be 8,000 or 100,000 years in the future. There are also signs that a global warming is in progress at this time.

One example of how items can be buried in Antarctic/Greenland snow is given by a P-38 airplane, which was abandoned on the snow in Greenland in 1942. It was dug out 50 years later in 1992, when it was 270 feet (82 meters) deep in the snow/ice. **At that location in Greenland, the average *snow/ice accumulation* (water equivalent) was 5.4 feet (1.6 meters) per year. From all of this, the experts *will not conclude* that the ice-mass in the whole Greenland is increasing.**

Google: **< p-38 airplane dug out of greenland ice in 1992 >.** Click on the link '*the lost squadron*'.

At the South Pole Station in Antarctica, the average *water equivalent of **annual precipitation*** has been 0.8 inches (2 centimeters) or approximately 5 to 7 inches (12 to 18 centimeters) of unpacked snow. The buildings get buried in snow 'down there'.

Google: < geothermal energy >

< volcanoes >

< heat capacity of air >

< earth's atmosphere >

< http://www.clearlight.com/~mhieb/WVFossils/ice_ages.html >

ISN'T THIS EARTH NEAT!

60. INDEX

A

Acceleration, 46, 49, 69, 87, 137, 144, 198, 200, 203, 208, 226, 232, 241, 252, 255, 270

Air, 13, 55, 72, 77, 100, 121, 130, 135, 161, 184, 204, 223, 258, 264, 296, 297

  density, 320

  air pressure, 121, 163, 270, 320

  speaking and hearing, 130

Andromeda Galaxy, 15, 33, 40, 45, 60

Antipode, 216, 248

Aphelion, 97, 149, 161, 220

Apocenter, 149

Apogee, 149

Archimedes, 166, 239, 301, 304, 307, 308

Arrow catchers, 108, 116

Artificial satellites, 25, 90, 100, 233

Asteroids, 17, 25, 31, 47, 69, 81, 91, 95, 100, 109, 113, 136

Astronomical unit, 13, 32, 35, 42, 47, 66, 82, 91, 106, 231, 268

Astronomy, 13, 21, 106, 159, 190, 195, 278

Atmosphere, 172, 179, 195, 202, 204, 226, 251, 258, 270, 286, 290, 294, 297, 314, 318, 320, 323

Atomic clocks, 28, 30, 183, 266, 276

Atoms, 15, 35, 129, 152, 208, 291

Autumnal Equinox, 193, 268

Axis, 49, 53, 59, 65, 69, 83, 97, 126, 158, 160, 163, 170, 185, 188, 193, 195, 199, 203, 228, 275

Azimuth, 23, 50, 176, 180

B

Balloon model, 54-55, 182, 197, 245, 292, 301

Barycenter, 70-71, 257

Bathymetric maps, 238, 244

Big bang, 15, 35, 43, 64, 152

Black hole, 65, 81, 85, 126, 227, 259

C

Celestial sphere, 37, 127, 177, 185, 187, 190, 215, 268

Centrifugal acceleration, 42, 49, 137, 158, 199, 203, 211, 212, 225, 229, 231, 243, 257, 308

Centrifuge, 87, 88, 137, 200, 234

Circumference, 66, 83, 159, 243, 250

Columbia Shuttle, 272

Comets, 14, 25, 31, 47, 69, 81, 89, 97, 105, 109, 113, 116, 134, 139, 147, 150, 168, 205, 219, 271, 287, 322

Compass directions, 25, 50, 176, 178, 219, 236, 243, 247

Constellations, 38, 40, 126, 170, 184, 191, 260, 293
Core, 13, 197, 230, 299, 301
Crust, 166, 253, 299, 301, 308

D

Deceleration, 56, 165, 198, 207, 318
Declination, 37, 187, 189, 216, 260, 282
Diurnal circle, 49, 185, 215, 261, 269
Doublet stars, 44, 144
Dust, 15, 72, 89, 94, 101, 111, 125, 137, 141, 272, 315, 320, 323

E

Earth, 13, 21, 24-25, 46, 65, 86, 155, 194, 248, 272, 281, 296, 310, 315
Earth tides, 162, 238, 240, 251-252, 254, 258, 304
Ecliptica, 14, 68, 103, 111, 117, 169, 187, 205, 259, 269, 281, 292
Ecliptic plane, 38, 70, 91, 103, 111, 163, 193, 250, 267, 290
Ellipse, 30, 53, 68, 90, 97, 104, 148, 159, 177, 219, 245, 257, 267, 282
Equation of time, 281-282
Equinox, 71, 189, 194, 268, 277
Equipotential surfaces, 236, 240, 248
Erathostenes, 19, 52, 243
Evolution, 15, 125, 132, 152-153, 273
Extra terrestrial, 44, 141
Extra-solar planets, 36, 124, 133, 143, 284

F

Flat Earth Society, 243, 247
Fundamental physical constants, 33, 69

G

Galaxies, 13, 15-17, 20, 32, 33, 37, 41, 42, 45, 47, 58-59, 60-61, 64-67, 74-75, 79, 125-127, 152, 162, 164, 174, 187-188, 190-191, 250, 261, 264-265, 278, 280
Galileo Galilei, 19, 71, 171, 227, 279
Geodesy, 22-23, 50, 52, 178, 233, 241
Geoid, 53, 159, 180, 187, 218, 220, 235, 238, 241, 254
Geoid undulations, 53, 180, 220, 235, 244
Global Positioning System, 56, 172, 179, 183, 277
Gravitation, 25, 42, 46, 53, 66, 69, 87, 90, 96, 101, 107, 113, 137, 148, 158, 162, 192, 197-198, 200, 202, 215, 226, 229, 231, 238, 241, 251, 265, 270, 287, 289, 298, 308, 316, 319
Gravity, 23, 25, 44, 54, 69, 78, 87, 88, 110, 137, 139, 166, 197, 198, 200, 203, 205, 208, 211-212, 215, 222, 225, 230, 241, 248, 252, 299, 306, 308, 314
Gyration, 69, 192

H

Heaven, 132, 262
Hell, 132, 262
Highest mountain, 49, 244
Holloman AFB, 209
Hydrostatic pressure, 254, 271
Hyperbola, 90, 101, 103, 147, 149, 271

I

Içe age, 165-166, 304-305, 326
Inclination, 71, 89, 101, 117, 163, 168, 272, 292

Inertia, 192, 195, 217, 241, 270
Intergalactic space, 96, 101

J

Jet Propulsion Laboratory, 17, 20, 269
Jupiter, 16, 41, 46, 78, 90, 96, 106, 117, 122, 127, 135, 141, 145, 161, 169, 248, 255, 286, 290, 293, 320

K

Kilimanjaro, 296
Kinetic energy, 98, 101, 110, 116, 144, 147, 161, 287

L

Latitude, 19, 25, 29, 39, 50, 65, 71, 86, 134, 158, 179, 186, 190, 196, 205, 212, 216, 221, 228, 236, 244, 250, 316
Law of buoyancy, 301-302
Law of inertia, 270
Leap second, 25, 160, 172, 269, 276-277
Leap years, 174, 278-279
Leonids, 99, 105, 110, 114
Length, 28-34, 62, 98, 105, 170, 173, 179, 195, 205, 220, 223, 243, 249, 270, 275, 281
Libration, 213, 214
Light-year, 13-15, 32, 34, 38, 40, 58, 61-67, 74, 78, 80, 121, 125-126, 144, 164, 191, 216, 259, 285, 297, 312
Light-second, 13-14, 31-35, 61, 66, 85, 109, 164, 169, 215, 231, 259, 267, 286, 294
Longitude, 25, 30, 38, 50, 160, 173, 179, 181, 186, 189, 233, 236, 244, 250, 276, 281
Lunation, 214

M

Mantle, 166, 197, 253, 255, 299-309
Mariana Trench, 49, 235, 243-244
Mars, 46, 78, 90-96, 106, 108, 116-117, 121, 131, 134, 136, 149, 161, 169, 217, 227, 240, 255, 262, 285, 289-290, 293, 321
Mercury, 14, 46, 77, 91, 96, 102, 107, 116, 119, 126, 133, 141, 161, 168, 204, 258, 269, 286, 287, 292
Meteorites, 25, 72, 89, 92, 98, 106-112, 121, 140, 146-147, 271, 274, 322
Meter, kilometer, 14, 28-30, 40, 45-46, 53, 58, 62, 66, 72, 85, 97, 113, 144, 158, 185, 197, 222, 238, 247, 270, 293, 300
Milky Way Galaxy, 13-17, 24, 32, 40, 65, 75, 81, 125, 162, 191, 260, 291
Molecule, 15, 35, 78, 93, 107, 129, 151, 227, 262, 271, 297, 302, 314
Moon, 25, 31, 47, 56, 63, 70, 149, 162, 168, 171, 191, 200, 203, 213, 232, 253, 266, 284, 294, 316, 320
Mount Everest, 49, 227, 245

N

NASA, 14, 16, 20, 55, 91, 106, 110, 122, 146, 159, 182, 214, 238, 241, 264, 295, 298, 318, 327
Nautical mile, 29, 236-237, 247
Neptune, 78, 123, 135, 141, 161, 169, 286, 290, 293
Nine solar planets, 31, 68-69, 77-78, 85, 90, 92-97, 104, 116, 120-126, 141, 168, 229, 267-270, 300, 320
NIST, 16, 28, 30, 170-171, 229, 279, 320

NOAA, 16, 20, 182, 190, 240, 243, 252, 254, 256, 258, 313

North Star, 38-39, 71, 103, 126, 185, 189, 193, 259

Nutation, 69, 188, 194

O

Ocean tides, 163, 253-258, 309

Obliquity of ecliptic plane, 72, 187-188, 282-283

Orbital speed, velocity, 42, 47, 53, 68, 93, 96-99, 107, 109, 113, 126, 161- 162, 174, 204, 219, 232, 245, 265, 269, 270, 282, 287, 289

Orbits, 13-15, 25, 37, 41-46, 53-56, 63, 68, 71, 81, 89-92, 97-103, 108, 117- 118, 120, 126, 134, 140, 146, 148- 150, 158, 160-162, 168, 173, 183, 187, 191-195, 203, 205, 214, 219-220, 226, 231, 252, 257, 259, 265-269, 272, 280, 287, 292, 318

P

Parabola, 91, 148, 219, 220

Pericenter, 149, 219

Perihelion, 47, 68, 90, 96, 102-103, 108, 118, 134, 140, 149, 161, 193, 269, 282

Pinwheel arms in Milky Way, 14, 32, 40, 64, 66, 94, 125, 259, 272

Planets, 12, 14, 17, 25, 31, 36, 40-46, 59, 68, 73-81, 85, 90, 95, 101, 107, 116, 119, 125-135, 141-143, 148, 161, 168, 187, 193, 195, 204, 241, 255, 269, 284, 287, 292, 321

Plumb line, 138, 180-186, 192, 198, 206, 215, 236-240, 255, 258, 262

Pluto, 14, 32, 37, 46, 63, 67, 78, 90, 104, 117, 122-128, 135, 141, 161, 168, 290, 292

Polaris, 38-39, 71, 103, 126, 178, 185, 188, 193, 194, 259

Precession, 69, 72, 189, 191-192, 195, 261

Proper motion, 126, 190, 191

Proxima Centauri, 14, 40, 144

Q

Quaoar, 107, 117, 134

R

Radian, 83, 249

Revolution, 52, 102, 159-160, 170, 180, 185, 214, 235, 243

Right ascension, 37, 42-43, 187-194, 261

Rotation, 23, 49, 53, 85, 102, 158-160, 170, 172, 185, 188-189, 192, 195, 198, 205, 213, 225, 235, 243, 251, 258, 269, 276, 278, 280, 308

S

Satellites, 20, 25, 52, 56, 89-90, 100, 108, 127, 134, 139, 148, 150, 160, 172, 177, 179, 182, 184, 219, 226, 233, 239, 256, 280, 318, 322, 327

Saturn, 46, 78, 99, 105, 108, 123, 127, 135, 141, 161, 169, 217, 248, 286, 290, 293, 320

Sea level, 23, 49, 51-53, 94, 137, 159, 180, 183, 186, 197, 201, 211, 218, 229, 232, 235, 238-244, 254, 264, 271, 287, 296-302, 305-312, 318-320

Sea level topography, 244

Second of time, 171, 247
Shooting stars, 72, 89, 93, 99, 109, 110
Sidereal day, 163, 173-175, 207, 277, 280
Solar day, 34, 173-175, 275-281
Solar system, 13-16, 20-31, 36, 41-43, 47-
48, 61, 65-73, 78, 81, 85, 89-93, 97,
100-107, 113, 116, 119, 124, 130-136,
139, 143, 147, 151, 155- 156, 162, 168,
260, 265, 269, 272, 284, 290, 292,
297, 313, 320, 323
Solar year, 34, 171, 173-175, 278, 280
Spheroid, 34, 53, 159, 213, 216, 236
Stationary objects, 46, 89, 99, 105, 160,
183, 213, 216, 265
Summer solstice, 269, 282
Sun, 13-25, 31, 35-40, 44-48, 53, 57, 61-
71, 75, 77-81, 85, 87, 90, 96-99, 103,
108, 116, 126, 134, 139, 146, 151,
155, 161, 173, 176, 187, 191, 203, 216,
227, 237, 241, 251, 269, 281, 290, 315,
326
Sundial, 280-281

T

Telescopes, 21, 37-38, 41, 65, 71, 84, 101,
127, 134, 178, 186-187, 190, 261, 279
Tides, 162-163, 238, 240-241, 251-258,
303, 309, 311
Time keeping, 25, 170, 172, 174, 275, 281
Topographic elevations, 25, 49-51, 159,
166, 179, 180-183, 206, 211-215, 221,
223, 229, 232-233, 238, 240, 244-
249, 296-297, 301-304, 309-311, 316,
319, 321
Topographic maps, 181, 240, 244
Travel times, 79, 143-144, 278, 285
Tsunami, 241, 254, 256

U

Unreachable stars, 36, 44, 124, 141, 321
Uranus, 46, 78, 123, 135, 141, 161, 169,
286, 290, 293
US Geological Survey, 20, 102, 244, 246,
300, 310

V

Vacuum of space, 46, 55, 72, 113, 139, 226,
245, 262, 267, 270, 294, 322-323
Velocities, 28, 34, 42, 46-47, 56, 66, 72,
92, 98, 100, 103, 109-110, 126, 144,
146, 160-163, 204, 208, 219, 232,
264-265, 270-273, 282, 287, 289, 322
Venus, 46, 77, 95, 108, 116, 119-120, 133,
136-137, 168, 262, 285, 288, 292
Vernal Equinox, 189, 192, 194, 268

W

Water, 44, 49, 77-78, 87, 89, 106, 119, 121-
122, 129, 131, 136, 139, 152, 162, 165,
172, 197, 201, 211, 213, 217, 225, 235,
239, 240, 244, 248, 253, 256, 271, 285,
288, 290, 294, 297-316, 325, 326
Winter Solstice, 269, 282

Z

Zenith, 185-186, 215-216, 218, 253